This book describes the physics of superconductivity and superfluidity, macroscopic quantum phenomena found in many conductors at low temperatures and in liquid helium 4 and helium 3.

In the first part of the book the author presents the mean field theory of generalised pair condensation. This is followed by a description of the properties of ordinary superconductors using BCS theory. The book then proceeds with expositions of strong coupling theory and the Ginzberg–Landau theory. The remarkable properties of superfluid helium 3 are then described as an example of a superfluid with internal degrees of freedom. The topics covered are dealt with in a coherent manner, with all necessary theoretical background given. Recent topics in the field, such as the copper oxide high temperature superconductors and exotic superconductivity of heavy fermion systems are discussed in the final chapter.

This book will be of interest to graduate students and researchers in condensed matter physics, especially those working in superconductivity and superfluidity.

T0215854

SUPERCONDUCTIVITY AND SUPERFLUIDITY

SUPERCONDUCTIVITY AND SUPERFLUIDITY

T. Tsuneto
Ryukoku University, Japan

Translated by Mikio Nakahara
Kinki University

CAMBRIDGE
UNIVERSITY PRESS

CAMBRIDGE UNIVERSITY PRESS
Cambridge, New York, Melbourne, Madrid, Cape Town, Singapore, São Paulo

Cambridge University Press
The Edinburgh Building, Cambridge CB2 2RU, UK

Published in the United States of America by Cambridge University Press, New York

www.cambridge.org
Information on this title: www.cambridge.org/9780521570732

Originally published in Japanese by Iwanami Shoten [1993]

English translation with revisions © Cambridge University Press 1998

This publication is in copyright. Subject to statutory exception
and to the provisions of relevant collective licensing agreements,
no reproduction of any part may take place without
the written permission of Cambridge University Press.

First published 1998
This digitally printed first paperback version 2005

A catalogue record for this publication is available from the British Library

Library of Congress Cataloguing in Publication data

Tsuneto, T. (Toshihiko), 1930–
Superconductivity and superfluidity / T. Tsuneto;
translated by Mikio Nakahara.
p. cm.
Includes bibliographical references and index.
ISBN 0 521 57073 5
1. Superconductivity. 2. Superfluidity. I. Title.
QC611.92.T75 1998
537.6′23–dc21 98-16451 CIP

ISBN-13 978-0-521-57073-2 hardback
ISBN-10 0-521-57073-5 hardback

ISBN-13 978-0-521-02093-0 paperback
ISBN-10 0-521-02093-X paperback

Contents

Preface

Superconductivity and superfluidity have already been studied for almost one century. In the case of superconductivity one may say it reached its peak with the dramatic appearance of BCS (Bardeen–Cooper–Schrieffer) theory in 1956. Since it was so successful, the research developed at an explosive rate, and by the beginning of the 1970s we thought the fundamental theories were well established. It is therefore logical that many of the well-known texts on superconductivity were written during this period. Superfluidity of liquid ^{4}He was also studied most actively during the early 1970s. Many people therefore had the impression that research in the field of superconductivity and superfluidity had already reached its peak. However, much interesting progress has been made since then. First, the superfluidity of liquid ^{3}He due to non-s-wave pairing was discovered in 1973, and became the subject of intense research, both experimental and theoretical. The generalised BCS theory again turned out to be quite successful, and over about ten years we obtained a basic understanding of this phenomenon. In the 1980s, stimulated by the superfluidity of liquid ^{3}He, people started to look for non-s-wave superconductivity in the heavy fermion systems. Then, rather unexpectedly, we witnessed the biggest event of recent years, namely the discovery of copper oxide high temperature superconductors by J.G. Bednorz and K.A. Müller in 1986 ([G-1]).

With this rich history before us, it is clearly an extremely difficult task to decide which topics to cover. A book of this size could readily be filled by a systematic treatise on superfluid ^{4}He alone. Indeed, a recently published book on superfluid ^{3}He consists of almost 600 pages. Hence we have selected the subject matter in a rather drastic manner. Since the high temperature superconductors (HTSC) are still attracting much attention, we have chosen superconductivity of fermion systems, including liquid ^{3}He, as the main subject of this book. As a consequence we treat the superfluidity of liquid ^{4}He only very briefly indeed, to our great regret.

Since the discovery of HTSC, enormous numbers of papers have been written on this and related subjects. This does not mean that the theory of superconductivity has become completely useless. Rather, in order to elucidate its mechanisms, it is necessary to have a thorough understanding of BCS theory and its generalisations such as non-s-wave pairing and the strong coupling theory. Therefore, a good deal of the book is devoted to these subjects. At the same time we must of course refer to major progress in the field made since the 1970s. As a result the style of the book may be rather compressed in some places. Even so, we had to omit important subjects such as macroscopic tunnelling, localisation and critical phenomena in superfluids and superconductors. Because of the limited space we had also to assume that readers possess basic knowledge of the theory of metals and the many-body theory. We hope that readers will make good use of the references listed at the end.

It is a great pleasure to thank Professors T. Ohmi and R. Ikeda for many illuminating discussions in the course of writing this book.

T. Tsuneto

Preface to the second edition

The recent observation of Bose–Einstein condensation in a polarised alkaline gas is of major significance to the content of this book. Although opinions may differ as to the significance of the observation itself to the physics, everybody must have applauded the impressive success achieved by the novel techniques. Thanks to the discovery which provided us with concrete examples of a dilute Bose gas discussed in Section 1.2, the macroscopic quantum wave function, the central concept in understanding superconductivity and superfluidity, now seems quite familiar to us. Although this subject has been treated in the Appendix of [E-11], we will briefly discuss it here from the viewpoint of the physics of the condensate. We hope that this is read together with the Appendix of [E-11].

The Nobel prize in physics for 1996 was awarded to D.D. Osheroff, R.C. Richardson and D.M. Lee for the discovery of superfluid ^{3}He. The research on superfluidity due to ^{3}P pairing, initiated by this discovery, taught us that the physics of pairing is quite rich. The lessons learned here have been helpful in research on high temperature superconductors, heavy electron systems and, moreover, superconductivity in hadronic matter.

One of the interesting developments in the research on HTSC is an attempt to observe directly the pairing type, now considered likely to be d. This will be discussed in Appendix A2.1. Another unique example of the physics of HTSC is its behaviour in a magnetic field, involving vortex lattice, vortex liquid and glass. These topics will be discussed in Appendix A2.2. The presentation may be more or less biased since even partial understanding of the system has become accessible only recently. Other recent progress will be considered in Appendix A2.3.

Professors K. Machida and K. Miyake have pointed out numerous errors in the first edition. Professors T. Ohmi, R. Ikeda and E. Maeno have made useful suggestions in the course of writing the appendices. The author would like to express his sincere thanks to all of them.

T. Tsuneto

Preface to the second edition

1
Superconductivity and superfluidity

A system composed of particles, whose quantum mechanical zero point energy is large compared to their interaction energy, does not solidify even at absolute zero temperature and remains a so-called quantum liquid. Typical examples of such systems are conduction electrons in metals and liquid helium. Many conductors undergo a phase transition at their critical temperature T_c and become superconducting below it. Similarly, liquid ^{4}He under its vapour pressure becomes a superfluid at $T_c = 2.17$ K (T_λ is often used instead of T_c), and liquid ^{3}He at $T_c = 0.9$ mK. Although the phenomena are called *superconductivity* for a charged system like conduction electrons, and *superfluidity* for a neutral system like liquid helium, they are characterised by the same basic property that the wave nature of the particles manifests itself on a macroscopic scale, as we shall see below.

1.1 Phenomena

The most important factor in determining the properties of a quantum liquid is the statistics of the constituent particles. Liquid ^{3}He becomes a typical Fermi liquid below 0.1 K and a superfluid below a few mK, in contrast with the bosonic liquid ^{4}He mentioned above. The two superfluids are quite different both phenomenologically and microscopically. Table 1.1 shows systems which exhibit, or are predicted to exhibit (in brackets), superconductivity or superfluidity. In the case of fermion systems pair formation is responsible for the phenomena so that their properties differ according to the pairing type.

Liquid ^{3}He and heavy fermion systems will be treated in Chapters 6 and 7, respectively. A neutron star is a stellar object whose density is greater than that of nuclei and the constituent neutrons and protons are thought to form independent superfluids. One can obtain spin-polarised hydrogen gas by collecting hydrogen atoms H$\downarrow$ with a polarised electron spin by

Table 1.1. *Systems exhibiting superconductivity or superfluidity*

Fermi-type; pair formation	Bose-type
Conduction electrons	Liquid ^{4}He
Many metals and alloys – ^{1}S	[Spin polarised hydrogen H$\downarrow$ gas]
Heavy fermion systems – ^{3}P, ^{1}D?	[Excitonic gas]
Liquid ^{3}He – ^{3}P	
[^{3}He in ^{4}He – ^{3}He mixture – ^{1}S?]	
[Nucleons in a neutron star – ^{1}S, ^{3}P]	

the use of a nonuniform magnetic field. These spin-aligned atoms do not form spin-singlet hydrogen molecules H_2. An exciton is a bound state of an electron and a hole formed by light absorption in an insulator, and may be considered as a Bose particle. Laser pulse irradiation of a Cu_2O crystal is used, for example, to obtain high density exciton gas.

^{1}S-superconductivity in ordinary metals and superfluidity in liquid ^{4}He are typical examples of Fermi-type and Bose-type superconductivity and superfluidity, respectively. Using these two examples, therefore, we will explain the basic features of the phenomena in the rest of this chapter.

1.1.1 Thermodynamical ordered phase

A superconducting or superfluid phase is a thermodynamical phase different from the normal state at $T > T_c$ and the phase transition between the two phases is accompanied by such phenomena as the specific heat anomaly. Figure 1.1 shows the temperature dependence of the specific heat. It should be noted that the specific heat increases near $T \simeq T_c$. Since entropy is released as the temperature is lowered through T_c the superconducting (superfluid) state has less entropy than the extrapolated normal state at the same temperature. In other words, the superconducting phase is more ordered than the normal phase. Next let us examine well-known phenomena to clarify what order characterises the superconducting or the superfluid phase. We will be mainly concerned with superconductors and give only supplementary comments on superfluids.

1.1.2 Zero resistivity

The electrical resistivity ρ vanishes, in other words the electrical conductivity $\sigma = \rho^{-1}$ tends to infinity, in a superconducting state as is seen from Fig. 1.2.

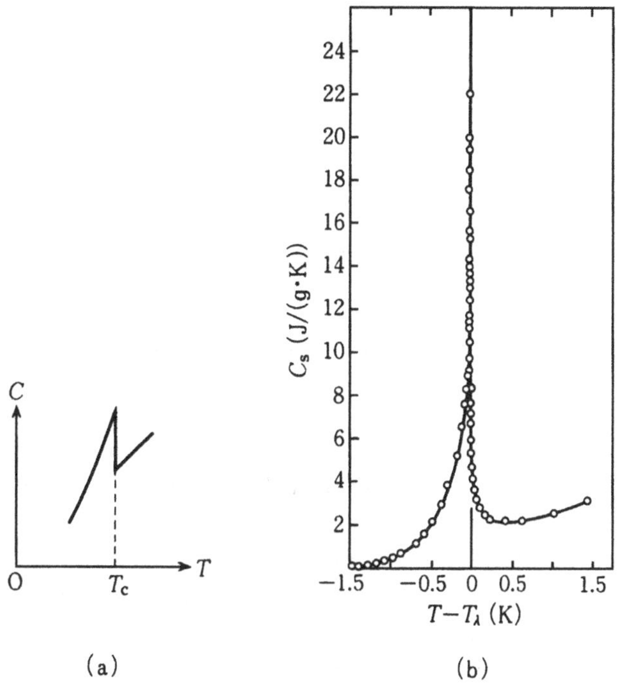

Fig. 1.1. The specific heat anomalies in (a) the superconducting transition and (b) the superfluid transition of ^{4}He. The specific heat of the latter tends to infinity as $|T - T_\lambda| \to 0$. See also Fig. 3.3 for the specific heat of a superconductor.

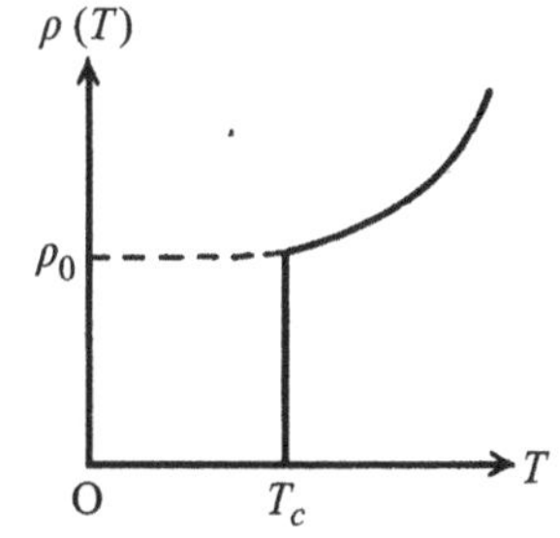

Fig. 1.2. The temperature dependence of the resistivity of a superconductor. The broken line shows the resistivity when the sample is kept in the normal state by applying a magnetic field H ($> H_c$) and ρ_0 is the residual resistivity in that case.

Historically, Onnes and his collaborators discovered superconductivity while they were measuring the temperature dependence of the resistivity of mercury at low temperatures (H. Kamerlingh Onnes (1911), see [C-7]). Even nowadays, it is this remarkable phenomenon that is measured first when we look for new superconducting materials. The analogous phenomenon for

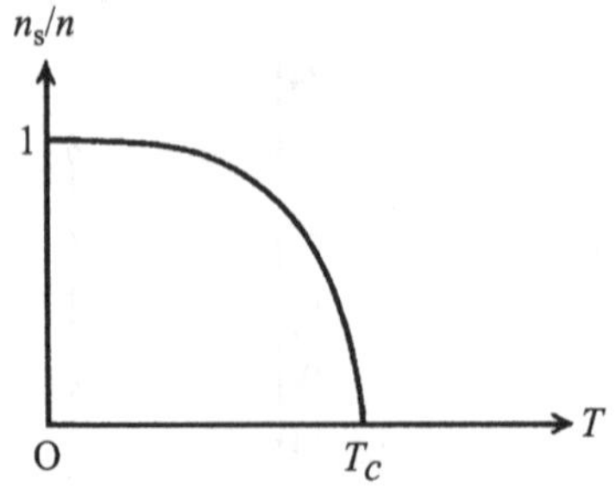

Fig. 1.3. The temperature dependence of the superfluid component n_s.

superfluid ^{4}He is flow without viscosity in a capillary tube with a radius of the order of 10^{-7} m. However, a nonvanishing viscosity is observed, if it is measured at finite temperatures through the decay of rotational vibration of a disk suspended in a fluid, as is commonly used to measure the viscosity of a liquid. The *two-component model* has been proposed based upon this observation and the so-called fountain effect. The fluid in this model is made of a superfluid component with number density n_s, which flows without viscosity and the normal component with density n_n, which has a finite viscosity and carries entropy. The component n_s takes a finite value at $T < T_c$ and $n_s = n$ (total number density) at $T = 0$ as shown in Fig. 1.3.

1.1.3 Perfect diamagnetism

If a rod of tin cooled below $T_c = 3.73$ K is placed in a weak magnetic field parallel to the rod, a diamagnetic surface current appears such that no magnetic field penetrates into the bulk sample. That is, it shows a perfect diamagnetism. This phenomenon is called the *Meissner effect.* If the external magnetic field H is raised above a temperature-dependent value $H_c(T)$, the sample ceases to be a superconductor and becomes a normal metal. The value $H_c(T)$ is called the *critical magnetic field.* The solid line in Fig. 1.4(a) shows the magnetisation M as a function of H. For some metals and alloys such as Nb_3Sn, however, the external magnetic field starts to penetrate into the sample at the *lower critical magnetic field* $H_{c1}(T)$ and the sample resumes its normal state above the *upper critical magnetic field* $H_{c2}(T)$, as shown by the solid line in Fig. 1.4(b). The former superconductor is called type 1, while the latter is called type 2. Figure 1.5 shows the phase diagram of a type 1 superconductor in the $H - T$ plane.

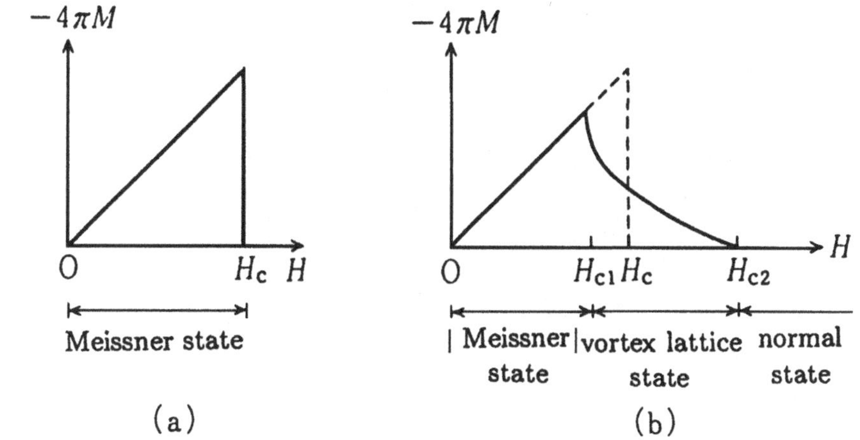

Fig. 1.4. Magnetisation in a superconducting state. (a) A type 1 superconductor and (b) a type 2 superconductor.

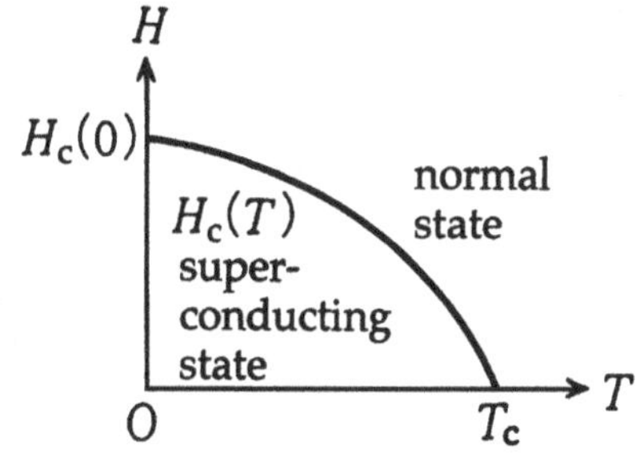

Fig. 1.5. The phase diagram of a type 1 superconductor.

1.1.4 Magnetic flux quantisation

Suppose a hollow cylinder made of a superconductor is cooled below its critical temperature T_c. One finds then that the magnetic field does not penetrate into the metal, while the flux passing through the cavity takes an integral multiple of the *flux quantum*

$$\phi_0 = 2\pi\hbar c/2e = 2.07 \times 10^{-15} \text{ weber } (= 10^{-7}\text{G} \cdot \text{cm}^2), \tag{1.1}$$

where $2\pi\hbar$ is the Planck constant, c is the velocity of light and e is the charge of an electron. Even when the magnetic field H is reduced to zero, the magnetic flux is still trapped in the cavity provided the metal is kept superconducting. As a consequence, there appears a persistent current along the inside surface of the cylinder (superconducting magnet). If an external magnetic field H such that $H_{c1} < H < H_{c2}$ is applied to a type 2 superconductor (see Section 1.1.3), the magnetic flux penetrates into the sample in

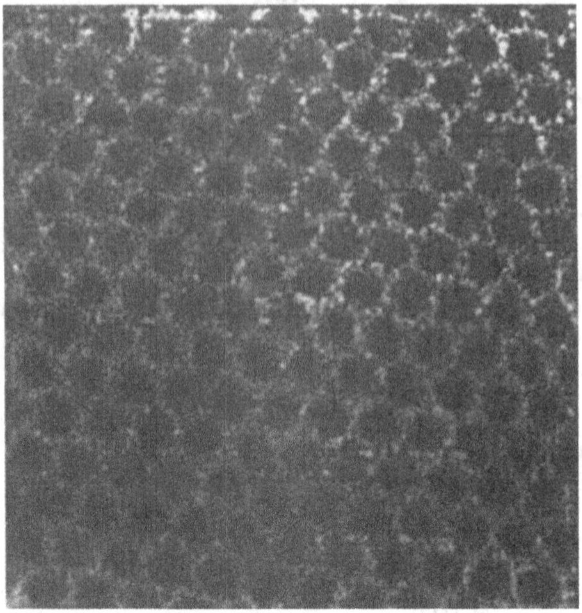

Fig. 1.6. Scanning tunnelling microscope image of a flux lattice. A magnetic field of 1 T amounts to a single magnetic flux per 489 × 489 Å^2. See also Section 5.4. (From [G-2].)

the form of a lattice as in Fig. 1.6 and a vortex current flows around each lattice cell. The magnetic flux through each lattice cell amounts to the flux quantum ϕ_0. The Josephson effect is also an inherent phenomenon in the superconducting state and will be discussed in Chapter 3.

The circulation in an electrically neutral superfluid ^{4}He is also quantised similar to the flux quantisation in a type 2 superconductor. Suppose a fluid has a velocity field $\boldsymbol{v}(\boldsymbol{x})$. The *circulation* along a closed loop C is a line integral $\Gamma = \oint_C \boldsymbol{v} \cdot d\boldsymbol{l}$. If a superfluid ^{4}He is rotated together with its vessel with a constant angular velocity ω, one finds a lattice of vortices parallel to the rotation axis. Then the circulation along a loop around a single vortex is quantised in the unit of a *circulation quantum*

$$\kappa \equiv 2\pi\hbar/m_4 = 0.997 \times 10^{-5}\ \mathrm{m^2\,s^{-1}} \tag{1.2}$$

where m_4 is the mass of a ^{4}He atom. There exist $2\omega/\kappa$ circulation quanta per unit area when the angular velocity is ω.

1.2 Bose gas and macroscopic wave function

In the remainder of this chapter, we will study a simple Bose system, while in the following chapters, we will be concerned mostly with Fermi systems. Let N be the total number of particles and V be the volume. An ideal noninteracting Bose gas undergoes a phase transition called the *Bose–Einstein condensation* at the temperature T_{BE} given by

$$k_{\mathrm{B}} T_{\mathrm{BE}} = \frac{2\pi\hbar^2}{m}\left(\frac{N}{2.6V}\right)^{2/3} . \tag{1.3}$$

Here k_{B} is the Boltzmann constant. At a temperature $T < T_{\mathrm{BE}}$, the lowest energy one-particle state $\psi_0(\boldsymbol{x})$, which is an eigenstate with $\boldsymbol{p} = 0$ in a uniform system, is occupied by

$$N_0(T) = N[1 - (T/T_{\mathrm{BE}})^{3/2}]$$

particles, hence $N_0(T)$ is of the order of N. The part of the system in the ψ_0 state is called the *condensate* since a macroscopic number ($N_0(T)$) of particles occupy a single state (in momentum space). Since entropy is carried by the remaining thermally excited particles, it is natural to consider the condensate a superfluid and the rest a normal fluid. There is a form of long range order in a state with a Bose–Einstein (BE) condensate. This is best seen in the density matrix written in terms of the creation and annihilation operators $\hat{\psi}^\dagger(\boldsymbol{x})$ and $\hat{\psi}(\boldsymbol{x})$ as

$$\rho(\boldsymbol{x}, \boldsymbol{x}') \equiv \langle \hat{\psi}(\boldsymbol{x})\hat{\psi}^\dagger(\boldsymbol{x}')\rangle = N_0\psi_0(\boldsymbol{x})\psi_0^*(\boldsymbol{x}') + g(\boldsymbol{x} - \boldsymbol{x}'), \tag{1.4}$$

where $\hat{}$ is introduced to distinguish operators from one-particle states. Here $\langle\cdots\rangle$ denotes the statistical average, and $g(\boldsymbol{x} - \boldsymbol{x}')$ is the short range correlation within the de Broglie wavelength at a temperature T. It is assumed that the wave function $\psi_0(\boldsymbol{x})$ of the condensate changes only over a macroscopic scale. In the high temperature phase there is only short range order. In the state with the BE condensate, $\rho(\boldsymbol{x}, \boldsymbol{x}')$ remains finite and of the order of the particle density, however large $|\boldsymbol{x} - \boldsymbol{x}'|$ is. Moreover, its dependence on $\boldsymbol{x}$ and $\boldsymbol{x}'$ is given, so to speak, independently by the wave function ψ_0. This implies that there exists a kind of long range order. Since it appears in the off-diagonal component of the density matrix, it is called the ODLRO (off-diagonal long range order). This should be compared with the crystalline order given by $\langle \hat{\psi}^\dagger(\boldsymbol{x})\hat{\psi}(\boldsymbol{x})\hat{\psi}^\dagger(\boldsymbol{x}')\hat{\psi}(\boldsymbol{x}')\rangle \sim n(\boldsymbol{x})n(\boldsymbol{x}')$, where $n(\boldsymbol{x})$ is the density variation with lattice periodicity.

If we substitute the mass of a ^{4}He atom into m and the density n of liquid ^{4}He into N/V in Eq. (1.3), we obtain $T_{\mathrm{BE}} = 3.13$ K, which is not very different from $T_\lambda = 2.17$ K for actual liquid ^{4}He. Also entropy at T_{BE}

is not much different from its value at T_λ for the actual liquid. For these reasons F. London has proposed that the superfluidity of ^{4}He is due to the BE condensation.

1.2.1 Dilute Bose gas

Let us show that one may explain the facts outlined in Section 1.1 qualitatively if one uses a macroscopic wave function corresponding to the condensate of an ideal Bose gas. To concentrate on the superfluid component our arguments will be limited to the case $T = 0$. To mimic real liquid ^{4}He, however, it is assumed that a weak repulsive interaction exists between particles. The Hamiltonian of the system in the second-quantised form is

$$H = \int \mathrm{d}\boldsymbol{x} \left\{ \frac{\hbar^2}{2m} \nabla \hat{\psi}^\dagger \nabla \hat{\psi} - \mu \hat{\psi}^\dagger \hat{\psi} \right\} + \frac{1}{2} \int \int \mathrm{d}\boldsymbol{x} \mathrm{d}\boldsymbol{x}' U(\boldsymbol{x} - \boldsymbol{x}') \hat{\psi}^\dagger(\boldsymbol{x}') \hat{\psi}(\boldsymbol{x}') \hat{\psi}^\dagger(\boldsymbol{x}) \hat{\psi}(\boldsymbol{x}), \tag{1.5}$$

where U is a weak repulsive interaction and μ is the chemical potential introduced so that the total particle number is N. The ground state of an ideal Bose gas is

$$|\Psi_{0N}\rangle = \frac{1}{\sqrt{N!}} (a_0^\dagger)^N |0\rangle \tag{1.6}$$

where $a_0^\dagger$ creates a particle in the state $\psi_0(\boldsymbol{x})$ of the condensate and $|0\rangle$ is the vacuum state. One may employ the same state as a first order approximation provided that the interaction is sufficiently weak. This yields the lowest order approximation to a dilute Bose gas. The state $\psi_0(\boldsymbol{x})$ must be determined by the Hartree approximation, namely the state is obtained by minimising the energy expectation value $E_0 = \langle \Psi_{0N} | H | \Psi_{0N} \rangle$. Let us take a potential $U(\boldsymbol{x}) = g\delta(\boldsymbol{x})$ for simplicity. We put $g = 4\pi a \hbar^2/m$ if $U(\boldsymbol{x})$ is to represent a hard core with radius a. Then we obtain

$$\frac{E_0}{N} = \int \mathrm{d}\boldsymbol{x} \left\{ \frac{\hbar^2}{2m} \nabla \psi_0^* \nabla \psi_0 - \mu |\psi_0|^2 \right\} + g \int \mathrm{d}\boldsymbol{x} |\psi_0|^4.$$

The wave function $\psi_0(\boldsymbol{x})$ is a complex number and ψ_0^* and ψ_0 are regarded as being independent of each other. Thus the wave function $\psi_0(\boldsymbol{x})$ is obtained from the condition $\delta E_0 / \delta \psi_0^* = 0$, that is

$$-\frac{\hbar^2}{2m} \nabla^2 \psi_0 - \mu \psi_0 + g |\psi_0|^2 \psi_0 = 0. \tag{1.7}$$

This is a nonlinear Schrödinger equation similar to the Ginzburg–Landau equation studied in Chapter 5. Since

$$|\psi_0|^2 = \mu/g = \bar{n}_s$$

in a uniform state, where the superfluid particle density $\bar{n}_s$ is given approximately by N/V in the present case, Eq. (1.7) can be rewritten as

$$-\frac{1}{2}\xi^2\nabla^2\psi_0 - \left(1 - \frac{|\psi_0|^2}{\bar{n}_s}\right)\psi_0 = 0. \tag{1.8}$$

Here we have introduced the healing distance

$$\xi = \hbar/\sqrt{mg\bar{n}_s}. \tag{1.9}$$

If $|\psi_0|^2$ deviates from $\bar{n}_s$ at a certain point, it recovers the value $\bar{n}_s$ in about the healing distance. The momentum density $\boldsymbol{g}(\boldsymbol{x}) = (\hbar/2\mathrm{i})\cdot\{\psi^*\nabla\psi - \nabla\psi^*\cdot\psi\}$ has an expectation value

$$\langle \boldsymbol{g}(\boldsymbol{x})\rangle = n_s(\boldsymbol{x})\hbar\nabla\phi \tag{1.10}$$

in the present state, where ψ_0 and ψ_0^* have been written in terms of the amplitude and the phase as

$$\psi_0(\boldsymbol{x}) = \sqrt{n_s(\boldsymbol{x})}\mathrm{e}^{\mathrm{i}\phi(\boldsymbol{x})}.$$

Therefore the velocity field $\boldsymbol{v}_s = \langle\boldsymbol{g}\rangle/mn_s$ is that of a potential flow whose potential is the phase of the macroscopic wave function and the circulation around a closed loop in the fluid is quantised as $p\kappa$, p being an integer and κ given by Eq. (1.2). The solution of Eq. (1.8) for a vortex line along the z-axis takes the form $\psi_0 = \sqrt{n_s(\boldsymbol{x})}\mathrm{e}^{\mathrm{i}p\theta}$ and the velocity field is given by $v_{s\theta} = p\kappa/r$. The density n_s approaches zero along the z-axis, where v_s diverges, and as a result a vortex has a core of radius $\sim \xi$. Quantised vortices in superfluid ^{4}He with $|p| = 1$ have been observed by various methods and it is known that $\xi \sim 1$ Å.

1.2.2 Charged Bose gas

We have considered a neutral Bose gas so far. Suppose now that a Bose gas has an electric charge e^*. (It is assumed, of course, that charges of an opposite sign are distributed uniformly in space and the screened Coulomb interaction between gas particles is taken into account in the definition of g.) Since particles couple with a magnetic field in this case, one makes the substitution

$$\left(\frac{\hbar}{\mathrm{i}}\right)\nabla\psi_0 \rightarrow \left(\frac{\hbar}{\mathrm{i}}\right)\left[\nabla - \frac{\mathrm{i}e^*}{\hbar c}\boldsymbol{A}\right]\psi_0,$$

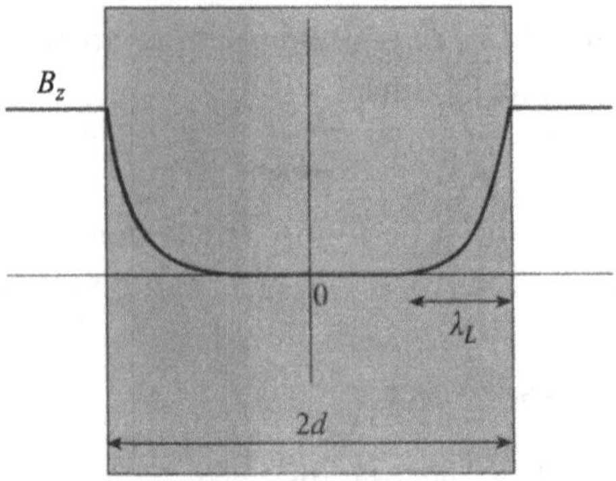

Fig. 1.7. Penetration of a magnetic field into a superconducting plate.

where $\boldsymbol{A}$ is the vector potential. Note in particular that the current density corresponding to $\langle \boldsymbol{g} \rangle$ is now

$$\langle \boldsymbol{j}_{\mathrm{s}} \rangle = \frac{e^* \hbar}{m^*} n_{\mathrm{s}} \left(\nabla \phi - \frac{e^*}{\hbar c} \boldsymbol{A} \right). \tag{1.11}$$

Where the scale of the spatial variation of the system is larger than ξ one may put $n_{\mathrm{s}} \simeq \bar{n}_{\mathrm{s}}$. Then, when combined with the Maxwell equation $\nabla \times \boldsymbol{B} = (4\pi/c)\langle \boldsymbol{j}_{\mathrm{s}} \rangle$, the following equation is obtained,

$$\begin{aligned} \nabla^2 \boldsymbol{B} - \lambda_{\mathrm{L}}^{-2} \boldsymbol{B} = 0, \\ \lambda_{\mathrm{L}}^{-2} \equiv 4\pi \bar{n}_{\mathrm{s}} e^{*2}/m^* c^2 \end{aligned} \tag{1.12}$$

where λ_{L} is called the *London penetration depth.* If the density $\sim 10^{23}\ \mathrm{cm}^{-3}$ of conduction electrons in a metal is substituted for $\bar{n}_{\mathrm{s}}$, and m^* and e^* are replaced by m and e of an electron, one obtains $\lambda_{\mathrm{L}} \sim 10^{-5}$ cm. Let us consider an actual example to see that Eq. (1.12) leads to the Meissner effect.

Suppose a superconducting plate of thickness $2d$ is placed in an external magnetic field H parallel to the plate. If Eq. (1.12) is solved in this case with the boundary conditions $B_z(\pm d) = H$, one obtains the solution

$$B_z(x) = H \frac{\cosh(x/\lambda_{\mathrm{L}})}{\cosh(d/\lambda_{\mathrm{L}})}, \quad (|x| < d).$$

Accordingly, for $d \gg \lambda_{\mathrm{L}}$, there is a current j_{s} along the $\pm y$-directions within a thickness λ_{L} of the surface of the plate and the magnetic field does not penetrate into the interior of the sample, see Fig. 1.7

Let us show next that this system leads to flux quantisation. Suppose a magnetic flux is trapped in the cavity of a superconducting cylinder as shown in Fig. 1.8. Since $\boldsymbol{B} = 0$ and $\langle \boldsymbol{j}_{\mathrm{s}} \rangle = 0$ along the dotted line C provided that

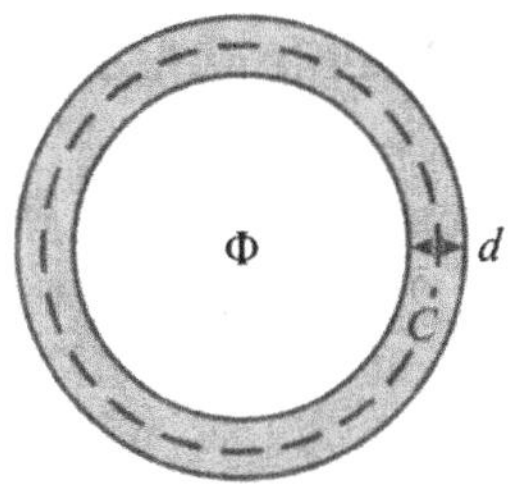

Fig. 1.8. A superconducting cylinder.

$d \gg \lambda_L$, it follows from Eq. (1.11) that

$$\oint_C \nabla\phi \mathrm{d}\boldsymbol{l} = \oint_C \frac{e^*}{\hbar c} \boldsymbol{A} \mathrm{d}\boldsymbol{l}.$$

By Stokes theorem, the right hand side of this equation is shown to be equal to $(e^*/\hbar c)\Phi$, where Φ is the total magnetic flux trapped inside the cylinder. Since ϕ in the left hand side is a phase, its total variation along C must be an integral multiple of 2π. Thus one is lead to

$$\Phi = 2\pi \frac{\hbar c}{e^*} p, \qquad (p = \text{integer}). \tag{1.13}$$

In fact, the observed flux quantisation in a superconductor is obtained if one puts $e^* = 2e$, which suggests that the superfluid consists of pairs of electrons. One reaches the identical conclusion even if an external magnetic flux exists only inside the cavity of the cylinder and the superconductor is not immersed in the field $\boldsymbol{B}$ directly. This is a good example of the *Aharonov–Bohm effect*, which states that a gauge field $\boldsymbol{A}$ itself has a physical meaning in quantum mechanics.

1.3 Symmetry breaking

The properties of a superfluid are best described by the wave function $\psi_0(\boldsymbol{x})$ as seen in the previous section. However, in a state with a condensate such as Eq. (1.6), one cannot regard the expectation value of $\hat{\psi}(\boldsymbol{x})$ as $\psi_0(\boldsymbol{x})$ since it vanishes provided the total number of particles is conserved. The density matrix of the form $\langle \Psi_{0N} | \hat{\psi}^\dagger(\boldsymbol{x}) \psi(\boldsymbol{x}') | \Psi_{0N} \rangle$ is rather inconvenient. Rather than consider the density matrix (1.4), we introduce the matrix elements

$$\langle \Psi_{0,N-1} | \hat{\psi}(\boldsymbol{x}) | \Psi_{0N} \rangle, \quad \langle \Psi_{0,N+1} | \hat{\psi}^\dagger(\boldsymbol{x}) | \Psi_{0N} \rangle. \tag{1.14}$$

One may consider, for the same purpose, expectation values with respect to the state

$$|\Psi_0\rangle = \frac{1}{\sqrt{2m+1}} \sum_{m'=-m}^{m} |\Psi_{0,N+m'}\rangle \tag{1.15}$$

which is an appropriate superposition of states with different N with $1 \ll m \ll N$. If N is a large number one has

$$\langle\Psi_0|\hat{\psi}(\boldsymbol{x})|\Psi_0\rangle = \psi_0(\boldsymbol{x}) \tag{1.16}$$

and the expectation value of various physical quantities with respect to this state is equal to that for a state with fixed N. It is this expectation value $\langle\hat{\psi}(\boldsymbol{x})\rangle$ in a state without fixed particle number that represents the long range order, characteristic of a superfluid state. The above expectation value is called the *order parameter*. It should be noted that the most familiar examples of the situation where particles occupy a single wave state, so that the wave function can be considered classical, are electromagnetic waves and laser lights. A typical example of a state such as Eq. (1.15) is an eigenstate of an annihilation operator called the *coherent state*.

Although a fermion system does not condense into a single-particle state due to the Pauli principle, there may appear an ODLRO in the two-particle density matrix $\langle\hat{\psi}_\alpha(\boldsymbol{x})\hat{\psi}_\beta(\boldsymbol{x}+\boldsymbol{r})\hat{\psi}^\dagger_{\beta'}(\boldsymbol{x}'+\boldsymbol{r}')\hat{\psi}^\dagger_{\alpha'}(\boldsymbol{x}')\rangle$ associated with the pair formation of particles. Here $\hat{\psi}^\dagger_\alpha(\boldsymbol{x})$ and $\hat{\psi}_\alpha(\boldsymbol{x})$ are creation and annihilation operators of particles with the spin state α. Accordingly the order parameter associated with superfluidity and superconductivity in a Fermi system is the pair wave function $\langle\hat{\psi}_\alpha(\boldsymbol{x})\hat{\psi}_\beta(\boldsymbol{x}+\boldsymbol{r})\rangle$, the explicit form of which will be analysed in detail in the following chapters.

A state with finite $\langle\hat{\psi}(\boldsymbol{x})\rangle$ or $\langle\hat{\psi}_\alpha(\boldsymbol{x})\hat{\psi}_\beta(\boldsymbol{x}+\boldsymbol{r})\rangle$ is not invariant under gauge transformations. Namely when one carries out a gauge transformation $U = \mathrm{e}^{\mathrm{i}\alpha\hat{N}}$ of the first kind, $\hat{N}$ being the number operator and α a constant, the change cannot simply be a constant phase multiplying the state vector

$$U|\Psi_N\rangle = \mathrm{e}^{\mathrm{i}\alpha N}|\Psi_N\rangle. \tag{1.17}$$

The expectation value itself changes as $\langle\hat{\psi}(\boldsymbol{x})\rangle \to \mathrm{e}^{\mathrm{i}\alpha}\langle\hat{\psi}(\boldsymbol{x})\rangle$. The Hamiltonian of a system may be left invariant under various transformations in general and the normal state observes this invariance. Ordering of a system implies that the system undergoes a transition to a state with less symmetries, that is, invariance under some transformations has been lost. The symmetry breaking associated with superconducting and superfluid transitions is gauge symmetry breaking. It should be noted here that if a system has an internal structure, such as the superfluid ^{3}He, the rotational symmetry is broken simultaneously.

It is the Josephson effect, explained in Section 3.4, that shows the theory dealing with $\langle\hat{\psi}\rangle$ or $\langle\hat{\psi}\hat{\psi}\rangle$ without fixed particle number to be essential in describing superconductivity and superfluidity. (The experiments are not possible for superfluids so far but one expects the same effects exist there.) The effect demonstrates that when two superconductors A and B are weakly coupled through a tunnelling junction so that electrons may be exchanged, the electron numbers N_{A} and N_{B} of respective superconductors do not take fixed values but the phase difference between the pair wave functions takes a definite value. In contrast to the case, for example, of a ferromagnet accompanied by a magnetic field, in superconductivity and superfluidity there is no symmetry breaking external physical field, that is, an external field that couples to the order parameter directly, To observe $\langle\hat{\psi}(\boldsymbol{x})\rangle$ 'directly', therefore, one has to couple the system with another similar system, such as a tunnelling junction, and see the response induced by the coupling.

1.4 Superfluidity

The particle number N_0 of the condensate is less than N in the presence of an interaction even in the ground state at $T = 0$. Suppose particles form a condensate in the $\boldsymbol{p} = 0$ state, there are processes in which two particles in the condensate may be scattered into states $(\boldsymbol{p}, -\boldsymbol{p})$ for example, which make the expectation value of the particle number $n_{\boldsymbol{p}}$ for a finite $\boldsymbol{p}$ nonvanishing. Neutron scattering experiments as well as theoretical estimates in liquid ^{4}He yield the result $N_0 \simeq 0.1N$. The ODLRO exists, however, since the density matrix is of the form (1.4) so long as N_0 is of the order of N. The $N - N_0$ particles in the present system must maintain a certain correlation with the condensate to lower the interaction energy. Therefore when the condensate flows with velocity $\boldsymbol{v}_{\mathrm{s}}$, the system should look like the ground state for $\boldsymbol{v}_{\mathrm{s}} = 0$ if seen by an observer moving with velocity $\boldsymbol{v}_{\mathrm{s}}$. Thus one reaches the conclusion that all the particles, not just N_0 particles, are thought of as the superfluid component. In other words, one has $n_{\mathrm{s}} = n$ at $T = 0$ K.

In reality, wall scattering occurs and it is a necessary condition for superfluidity that the flow does not decay by these scattering processes. In order to discuss this one has to know not only the ground state but the excited states. Suppose an elementary excitation with a momentum $\boldsymbol{p}$ has an energy $\varepsilon_{\boldsymbol{p}}$. If the superfluid moves with a velocity $\boldsymbol{v}_{\mathrm{s}}$, an elementary excitation with momentum $\boldsymbol{p}$ in the co-moving frame has the energy $\varepsilon_{\boldsymbol{p}}$ as mentioned above. The elementary excitation energy becomes $\varepsilon_{\boldsymbol{p}} + \boldsymbol{p} \cdot \boldsymbol{v}_{\mathrm{s}}$ if one Galilei-transforms

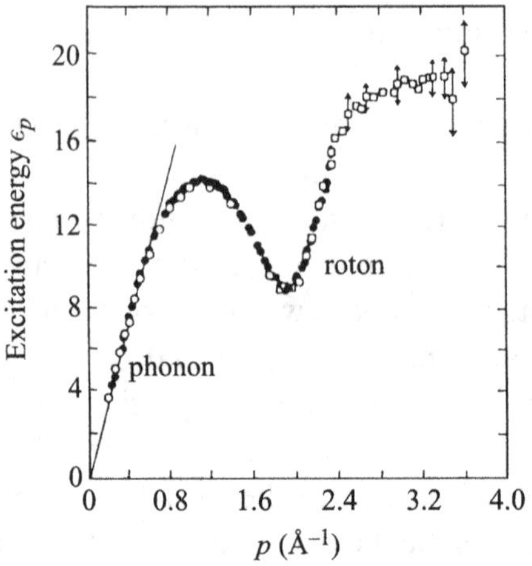

Fig. 1.9. Elementary excitations in superfluid ^{4}He measured by neutron inelastic scattering. The interval $p < 0.8$ Å^{-1} is called the phonon region while 1.4 Å^{-1} $<$ $p < 2.4$ Å^{-1} is called the roton region. (From [G-3].)

to the rest frame of the 'wall'. Therefore all the excitations should satisfy the *Landau criterion*

$$\varepsilon_{\boldsymbol{p}} + \boldsymbol{p} \cdot \boldsymbol{v}_s > 0 \tag{1.18}$$

for the flow energy not to be lowered by creating excitations by scattering off a wall, that is, for the superflow to be stable. An ideal Bose gas does not satisfy this condition. Elementary excitations in liquid ^{4}He are phonons and rotons with the Landau spectrum shown in Fig. 1.9 and satisfy the above condition. The critical velocity $\boldsymbol{v}_{sc}$ of a superflow is the minimal $\boldsymbol{v}_s$ for which an equality in Eq. (1.18) is realised for a certain excitation. The critical velocity in the superfluid ^{4}He is determined in many cases not by the phonon or roton excitations but by quantised vortices.

In the presence of an interaction, particle-like excitations with the energy spectrum $\boldsymbol{p}^2/2m$ in an ideal Bose gas are absorbed into a collective mode called the sound wave (density wave) with the spectrum pv_s. It was shown explicitly by the theory of an interacting dilute Bose gas (N.N. Bogoliubov, 1947, see [F-2]) that the existence of a condensate is indispensable for this mechanism. Bogoliubov's theory starts with the mean field associated with the condensate. The Bogoliubov transformation employed in the theory of superconductivity first appeared in this context. In a dilute Bose gas considered in the previous section a fluctuation with wave number p around the condensate $\psi_0 = \sqrt{\bar{n}_s}$ has a spectrum of the phonon type. In order to see this, one considers the equation of motion corresponding to Eq. (1.7)

$$\left(\frac{\mathrm{i}}{\mu}\frac{\partial}{\partial t}+1\right)\psi=-\frac{1}{2}\xi^2\nabla^2\psi+\frac{1}{\bar{n}_\mathrm{s}}|\psi|^2\psi,$$

substitutes $\psi=\sqrt{\bar{n}_\mathrm{s}}+\delta\psi$ and then linearises the resulting equation in $\delta\psi$. One finds as a result the eigenfrequency

$$\omega^2=(\bar{n}_\mathrm{s}g/m)p^2+(p^2/2m)^2$$

which becomes a sound wave with dispersion $v_\mathrm{s}=\sqrt{\bar{n}_\mathrm{s}g/m}$ as $p\to 0$.

At finite temperature there are phonon and roton excitations according to the equilibrium distribution in the rest frame of a 'wall'. Landau constructed the theory of superfluid ^{4}He from the viewpoint that these thermal excitations constitute the normal component, and successfully treated the thermodynamic properties and transport phenomena quantitatively. We also note R.P. Feynman's unique theory concerning the phonon and roton spectrum of the elementary excitations in superfluid ^{4}He (see [F-2]).

2

Mean-field theory of pair condensation

The microscopic theory of superconductivity of metals was discovered by J. Bardeen, L.N. Cooper, and J.R. Schrieffer (BCS) in 1957. The key concept of the BCS theory is pair condensation. In the present chapter, the mean-field theory of superfluidity and superconductivity is developed for ordinary superconductors of singlet s-wave pairing and superfluid ^{3}He with triplet p-wave pairing. Our formalism is applicable to such systems as neutron stars and heavy fermion superconductors where pairs have internal degrees of freedom.

2.1 Interactions and two-particle bound states

Bound state formation by two particles on the Fermi surface is called *pairing*. Before we discuss pairing in many-body systems, let us consider interaction between two fermions with spin 1/2 and their bound states.

Suppose there exists an interaction H_{I} between two particles, whose matrix element is given by

$$\langle \boldsymbol{k}_+\alpha, -\boldsymbol{k}_-\beta | H_{\mathrm{I}} | \boldsymbol{k}'_+\alpha', -\boldsymbol{k}'_-\beta' \rangle \equiv V_{\alpha\beta,\alpha'\beta'}(\boldsymbol{k}, \boldsymbol{k}'), \tag{2.1}$$

where $|\boldsymbol{k}_+\alpha, -\boldsymbol{k}_-\beta\rangle$ denotes a state vector of two fermions with momenta $\boldsymbol{k}_+ = \boldsymbol{k} + \boldsymbol{q}/2$ and $-\boldsymbol{k}_- = -\boldsymbol{k} + \boldsymbol{q}/2$ and spins α and β. This vector is also written in the second-quantised form as

$$|\boldsymbol{k}_+\alpha, -\boldsymbol{k}_-\beta\rangle = a^\dagger_{k_+\alpha} a^\dagger_{-k_-\beta} |0\rangle$$

where $|0\rangle$ indicates vacuum[1]. $a^\dagger_{k\alpha}$ is an operator which creates a particle of momentum $\boldsymbol{k}$ and spin α. It is assumed, for simplicity, that the matrix

[1] Vectorial subscripts will be denoted by ordinary italic letters hereafter if there is no risk of confusion. Accordingly $a_{\boldsymbol{k}}$ is simply denoted by a_k.

element (2.1) is independent of $\boldsymbol{q}$. It follows from the antisymmetry of fermionic operators that

$$V_{\beta\alpha,\alpha'\beta'}(-\boldsymbol{k},\boldsymbol{k}') = V_{\alpha\beta,\beta'\alpha'}(\boldsymbol{k},-\boldsymbol{k}') = -V_{\alpha\beta,\alpha'\beta'}(\boldsymbol{k},\boldsymbol{k}'). \tag{2.2}$$

The matrix elements also satisfy the Hermiticity condition

$$V^*_{\alpha\beta,\alpha'\beta'}(\boldsymbol{k},\boldsymbol{k}') = V_{\alpha'\beta',\alpha\beta}(\boldsymbol{k}',\boldsymbol{k}). \tag{2.3}$$

Suppose the interaction H_{I} leaves the spin invariant. Then it follows from Eq. (2.2) that

$$V_{\alpha\beta,\alpha'\beta'}(\boldsymbol{k},\boldsymbol{k}') = \frac{1}{2}\left(\delta_{\alpha\alpha'}\delta_{\beta\beta'} - \delta_{\alpha\beta'}\delta_{\beta\alpha'}\right) V^{(\mathrm{e})}(\boldsymbol{k},\boldsymbol{k}') + \frac{1}{2}\left(\delta_{\alpha\alpha'}\delta_{\beta\beta'} + \delta_{\alpha\beta'}\delta_{\beta\alpha'}\right) V^{(\mathrm{o})}(\boldsymbol{k},\boldsymbol{k}'). \tag{2.4}$$

Here

$$V^{(\mathrm{e}),(\mathrm{o})}(-\boldsymbol{k},\boldsymbol{k}') = V^{(\mathrm{e}),(\mathrm{o})}(\boldsymbol{k},-\boldsymbol{k}') = \pm V^{(\mathrm{e}),(\mathrm{o})}(\boldsymbol{k},\boldsymbol{k}'), \tag{2.5}$$

where (e) and (o) represent the even and the odd part of the interaction, respectively. If the interaction Hamiltonian is given by

$$H_{\mathrm{I}} = \frac{1}{2}\int V(|\boldsymbol{x}-\boldsymbol{x}'|)n(\boldsymbol{x})n(\boldsymbol{x}')\mathrm{d}\boldsymbol{x}\mathrm{d}\boldsymbol{x}',$$

for example, one obtains

$$V^{(\mathrm{e}),(\mathrm{o})}(\boldsymbol{k},\boldsymbol{k}') = \frac{1}{2}\left[V(|\boldsymbol{k}-\boldsymbol{k}'|) \pm V(|\boldsymbol{k}+\boldsymbol{k}'|)\right]. \tag{2.6}$$

Now let us consider a two-particle system. The energy eigenstate with centre of mass momentum $\boldsymbol{q}$ can be written as

$$\sum_{\boldsymbol{k},\alpha\beta} C_{\alpha\beta}(\boldsymbol{k},\boldsymbol{q})|\boldsymbol{k}_+\alpha, -\boldsymbol{k}_-\beta\rangle.$$

The coefficient $C_{\alpha\beta}(\boldsymbol{k},\boldsymbol{q})$ is the wave function in the momentum representation and satisfies the antisymmetry

$$C_{\beta\alpha}(-\boldsymbol{k},\boldsymbol{q}) = -C_{\alpha\beta}(\boldsymbol{k},\boldsymbol{q}). \tag{2.7}$$

The Schrödinger equation in this representation is

$$\left\{\frac{1}{2m}(\boldsymbol{k}_+^2 + \boldsymbol{k}_-^2) - E\right\} C_{\alpha\beta}(\boldsymbol{k},\boldsymbol{q}) + \sum_{\boldsymbol{k}',\alpha'\beta'} V_{\alpha\beta,\alpha'\beta'}(\boldsymbol{k},\boldsymbol{k}')C_{\alpha'\beta'}(\boldsymbol{k}',\boldsymbol{q}) = 0. \tag{2.8}$$

When the interaction is spin-independent and hence satisfies Eq. (2.4) it immediately follows that there are two classes of eigenfunctions, namely the spin-singlet state satisfying

$$C_{\alpha\beta}(\boldsymbol{k},\boldsymbol{q}) = C_{\alpha\beta}(-\boldsymbol{k},\boldsymbol{q}) = -C_{\beta\alpha}(\boldsymbol{k},\boldsymbol{q}) \tag{2.9}$$

and the spin-triplet state satisfying

$$C_{\alpha\beta}(\boldsymbol{k},\boldsymbol{q}) = -C_{\alpha\beta}(-\boldsymbol{k},\boldsymbol{q}) = C_{\beta\alpha}(\boldsymbol{k},\boldsymbol{q}). \tag{2.10}$$

The spin indices and the symmetry indices (e) and (o) will be omitted and $\boldsymbol{q} = 0$ will be taken, for simplicity, in the following. Then Eq. (2.8) is rewritten as

$$\Lambda(\boldsymbol{k}) \equiv \sum_{k'} V(\boldsymbol{k},\boldsymbol{k}')C(\boldsymbol{k}')$$

$$\Lambda(\boldsymbol{k}) = \sum_{k'} V(\boldsymbol{k},\boldsymbol{k}')\frac{1}{E - \boldsymbol{k}'^2/m}\Lambda(\boldsymbol{k}'). \tag{2.11}$$

Suppose the interaction is invariant under rotations in momentum space. Then the matrix element can be expanded as

$$\begin{aligned} V(\boldsymbol{k},\boldsymbol{k}') = V(k,k',\hat{\boldsymbol{k}}\cdot\hat{\boldsymbol{k}}') &= \sum_l (2l+1)V_l(k,k')P_l(\hat{\boldsymbol{k}}\cdot\hat{\boldsymbol{k}}') \\ &= 4\pi\sum_l V_l(k,k')\sum_{m=-l}^{l} Y_l^m(\Omega_{\hat{k}})Y_l^{-m}(\Omega_{\hat{k}'}), \end{aligned} \tag{2.12}$$

where $\hat{\boldsymbol{k}} \equiv \boldsymbol{k}/|\boldsymbol{k}|$ and $\Omega_{\hat{k}}$ is the solid angle. In this case, Eq. (2.11) can be written in terms of the partial wave component as

$$\Lambda_l^m(k) = \frac{1}{2\pi^2}\int_0^\infty dk'k'^2 V_l(k,k')\frac{1}{E - k'^2/m}\Lambda_l^m(k') \tag{2.13}$$

where $\Lambda(\boldsymbol{k})$ has been decomposed as

$$\Lambda(\boldsymbol{k}) = \sum_{l,m}\Lambda_l^m(k)Y_l^m(\Omega_{\hat{k}}).$$

A bound state corresponds to the solution with $E < 0$. It is not easy to find the solution to Eq. (2.13) due to a generally nontrivial dependence of $V(k,k')$ on k and k'. It can be shown, however, that if V_l is negative, bound states with an angular momentum l may exist. It can be seen from Eqs. (2.8) and (2.9) that a bound state with l even (odd) must be a spin-singlet (triplet) state. For example, the potential of a hydrogen atom is $V_{kk'} = 4\pi e^2/|\boldsymbol{k} - \boldsymbol{k}'|^2$ and the 1s-ground state wave function is given by (see [G-4])

$$C_k = \sqrt{\frac{2}{\pi}}\frac{4k}{(1+k^2)^2}.$$

Let us take an s-wave ($l = 0$) interaction of a δ-function type such that the potential is given by $V_0(k, k') \simeq -\lambda = $ const. for $k, k' \leq K_c$ and $= 0$ otherwise. For an attractive potential ($\lambda > 0$), solutions with $E < 0$ are determined by the condition

$$\frac{K_c}{\sqrt{m|E|}} = \tan \frac{K_c}{\sqrt{m|E|}} \left(1 - \frac{2\pi^2}{mK_c\lambda}\right). \tag{2.14}$$

The above equation also gives the bound state condition for a three-dimensional square well potential of depth U and width a if the parameters are chosen to be $\lambda = Ua^3$ and $K_c = a^{-1}$. Note that there are no bound states unless $4\pi m K_c \lambda > 1$. It should be also noted that Eq. (2.11) always has a bound state solution for a one-dimensional problem however small λ may be.

2.2 Cooper pairs and the BCS ground state

Due to the Pauli principle, not only an ideal Fermi gas, but an interacting Fermi system such as electrons in a metal and liquid ^{3}He, can, at low enough temperatures, be regarded as a system of quasi-particles obeying Fermi statistics; called the Fermi liquid. If there are no phase transitions, the ground state at $T = 0$ K is a Fermi sphere of quasi-particles. Let $a^\dagger_{k\alpha}$ be an operator which creates a quasi-particle (which we simply call a particle hereafter) of momentum $\boldsymbol{k}$ and spin α. Then the Fermi sphere is written as

$$|\Psi_F\rangle = \prod_{k,\alpha} a^\dagger_{k\alpha}|0\rangle, \tag{2.15}$$

where $|0\rangle$ is the vacuum state and the product is over the states inside the Fermi sphere. Let us recall that the number density n is given, in terms of the Fermi momentum k_F, as $n = k_F^3/3\pi^2$ and that the density of states on the Fermi surface, namely the number of one-particle states per unit volume and unit energy, is

$$\frac{\mathrm{d}n(k_F)}{\mathrm{d}(k_F^2/2m)} = \frac{m}{\pi^2}k_F \equiv 2N(0), \tag{2.16}$$

where $N(0)$ is the density of states for a fixed spin direction. The effective mass of the quasi-particle has been written simply as m. The importance in the physics of a Fermi system that all the states on the Fermi surface are degenerate should also be stressed. Let

$$H = \sum_{k,\alpha} \xi_k a^\dagger_{k\alpha} a_{k\alpha} + \frac{1}{2} \sum_{\substack{k,k',q \\ \alpha\beta,\alpha'\beta'}} V_{\alpha\beta,\alpha'\beta'}(\boldsymbol{k}, \boldsymbol{k}') a^\dagger_{-k_-\beta} a^\dagger_{k_+\alpha} a_{k'_+\alpha'} a_{-k'_-\beta'} \tag{2.17}$$

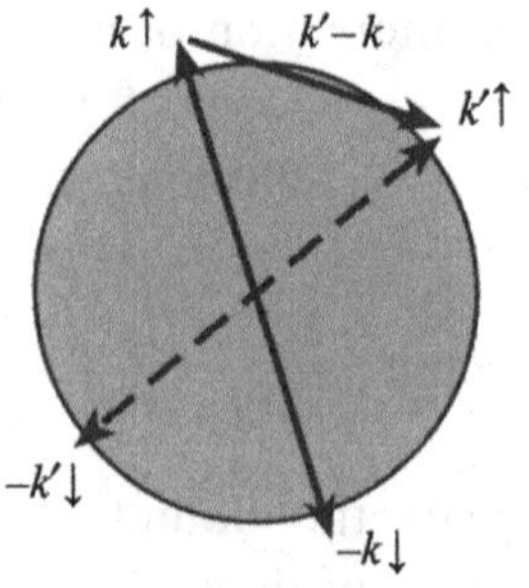

Fig. 2.1.

be the Hamiltonian of the system. Here the matrix element V is given by Eq. (2.1) and, if μ denotes the chemical potential,

$$\xi_k = \boldsymbol{k}^2/2m - \mu$$

is the particle energy measured from the Fermi surface. One may put $\mu = k_F^2/2m$ in an equilibrium state.

Now let us show that where the interaction is negative, however small it may be, there exists a state whose energy is lower than that of the Fermi sphere ground state. As an example, consider a state with a spin-singlet pair with centre of mass momentum zero outside the Fermi sphere (Fig. 2.1),

$$|\Psi\rangle = \sum_{k>k_F} C_k a_{k\uparrow}^\dagger a_{-k\downarrow}^\dagger |\Psi_F\rangle, \tag{2.18}$$

where $C_{\uparrow\downarrow}(\boldsymbol{k}) = -C_{\downarrow\uparrow}(\boldsymbol{k}) \equiv C_k$. Let us consider an approximation in which the only rôle of the interaction is to scatter the pair $(\boldsymbol{k}\uparrow, -\boldsymbol{k}\downarrow)$ into another pair $(\boldsymbol{k}'\uparrow, -\boldsymbol{k}'\downarrow)$, both pairs being outside the Fermi sphere. Then this problem becomes analogous to the two-particle problem studied in the previous section, except that the present pair is interacting in the presence of the background Fermi sphere and hence the k'-integration in (2.13) is over the range $k' > k_F$. Let us take an s-wave ($l = 0$) interaction of the form

$$V_{l=0}(k, k') = \begin{cases} -g & (\omega_c > \xi_k, \xi_{k'} > 0) \\ 0 & (\text{otherwise}). \end{cases} \tag{2.19}$$

where it is assumed that $\omega_c \ll \varepsilon_F = k_F^2/2m$. In other words, there exists a constant attractive potential between two particles lying within the range of ω_c from the Fermi surface. If one substitutes the $\boldsymbol{k}'$-summation by the integral

$N(0)\int d\xi$, one obtains an equation which determines the eigenvalue $-|E|$,

$$1 = N(0)g\int_0^{\omega_c} d\xi \frac{1}{|E| + 2\xi} = N(0)g \ln \frac{|E| + 2\omega_c}{|E|}. \qquad (2.20)$$

Note that there always exists a solution $|E|$ to this equation. Since the Fermi sphere ground state gives $E = 0$, the result above implies that the Fermi sphere state is unstable against the formation of the pair introduced above however small the attractive interaction may be and hence that the Fermi sphere state cannot be the true ground state. This fermion pair is called a *Cooper pair*. For $N(0)g \ll 1$ one finds from Eq. (2.20) that

$$|E| \simeq 2\omega_c e^{-1/N(0)g}. \qquad (2.21)$$

The singular dependence of $|E|$ on g indicates that this result cannot be obtained by a perturbative expansion with respect to g.

The wave function of a Cooper pair $C_k \propto (|E| + 2\xi_k)^{-1}$ in the k-representation corresponds to the real space wave function

$$\Psi(\boldsymbol{x}) = \sum_k C_k e^{i\boldsymbol{k}\cdot\boldsymbol{x}} \propto \int dk \frac{\sin kr}{kr} \frac{1}{|E| + 2(k - k_F)v_F}, \quad (k_F < k < k_c) \qquad (2.22)$$

where $\boldsymbol{x}$ is the relative coordinate of the pair, $r = |\boldsymbol{x}|$, and k_c is fixed by the relation $\xi_{k_c} \simeq (k_c - k_F)v_F = \omega_c$. The function $\Psi(r)$ varies as $r^{-1}\cos k_F r$ for small r and approaches zero faster than r^{-1} for $r \gtrsim v_F/|E|$. Accordingly one can take the size of the pair to be roughly $v_F/|E|$. Since a Cooper pair is formed above the Fermi sphere, the size of the pair is larger than that of a pair of the same binding energy in vacuum by a factor $\sqrt{\varepsilon_F/|E|}$.

2.2.1 The BCS ground state

When a Cooper pair exists due to an attractive interaction, it is expected that all the particles in the vicinity of the Fermi surface should form pairs of some kind to lower the total energy. To make our discussion simple, let us assume for the moment that the Cooper pair is a spin singlet as before. Then the pair wave function in the k-representation is given by $C_{\uparrow\downarrow}(\boldsymbol{k},\boldsymbol{q}) = C_{\uparrow,\downarrow}(-\boldsymbol{k},\boldsymbol{q}) \equiv C_k$, $C_{\uparrow\uparrow} = C_{\downarrow\downarrow} = 0$, see Eq. (2.9). It would be natural to expect that the particles all form identical pairs with the largest binding energy to minimise the system energy, just like the Bose–Einstein condensation in a Bose system.

It should be noted in particular that the centre of mass momentum $\boldsymbol{q}$ must be the same for all the pairs, to take advantage of the attractive matrix elements. Let us take $\boldsymbol{q} = 0$ here. Then the ground state is given by

$$|\Psi_N\rangle = \left(\sum_k C_k a^\dagger_{k\uparrow} a^\dagger_{-k\downarrow}\right)^{N/2} |0\rangle, \tag{2.23}$$

where N is the total number of particles. Written in the coordinate representation, this becomes a product of two-body wave functions:

$$|\Psi_N\rangle = \mathscr{A} \prod_{i \neq j} \Psi(\boldsymbol{x}_i - \boldsymbol{x}_j) s_{ij},$$

where $\mathscr{A}$ is the antisymmetrisation operator and s_{ij} denotes that the pair i, j is in the spin-singlet state. Equation (2.23) being a product of $N/2$ pairs of $a^\dagger_{k\uparrow} a^\dagger_{-k\downarrow}$, the wave function is a superposition of a state with a pair and a state without a pair, for any particular $\boldsymbol{k}$. Accordingly, $|\Psi_N\rangle$ can also be written as

$$|\Psi_N\rangle = P_N \prod_k \left(1 + C_k a^\dagger_{k\uparrow} a^\dagger_{-k\downarrow}\right) |0\rangle,$$

where P_N is the projection operator onto the states with N particles. Let us normalise the wave function as

$$|\Psi_N\rangle = P_N \prod_k \left(u_k + v_k a^\dagger_{k\uparrow} a^\dagger_{-k\downarrow}\right) |0\rangle = P_N |\Psi_0\rangle. \tag{2.24}$$

Here the coefficients u_k and v_k are given by

$$u_k = \left(1 + |C_k|^2\right)^{-1/2}, \quad v_k = C_k \left(1 + |C_k|^2\right)^{-1/2}. \tag{2.25}$$

The condition that the total particle number be N is not imposed on $|\Psi_0\rangle$. For this wave function one has

$$\bar{N} = \langle \Psi_0 | \sum_{k,\sigma} a^\dagger_{k\sigma} a_{k\sigma} | \Psi_0 \rangle = 2 \sum_k |v_k|^2$$

$$\overline{(N - \bar{N})^2} = 4 \sum_k |v_k|^2 \left(1 - |v_k|^2\right) = O(\bar{N}).$$

Therefore $|\Psi_0\rangle$ may be employed instead of $|\Psi_N\rangle$ for large N provided that $\bar{N} = N$ is satisfied. This approximate ground state $|\Psi_0\rangle$ was first introduced in the BCS theory. We note *en passant* that $|\Psi_0\rangle$ becomes the Fermi sphere $|\Psi_F\rangle$ if one puts $v_k = 1$, $(k \leq k_F)$ and $v_k = 0$, $(k > k_F)$. We will see in the next section that, in a superconducting state, $|v_k|$ takes a value between 0 and 1 in the vicinity of the Fermi surface, where the attractive interaction is at work.

The expectation values $\langle \Psi_0 | a_{-k\downarrow} a_{k\uparrow} | \Psi_0 \rangle$ and $\langle \Psi_0 | a^\dagger_{k\uparrow} a^\dagger_{-k\downarrow} | \Psi_0 \rangle$ do not vanish in the state $|\Psi_0\rangle$. Therefore, as remarked in Section 1.3, the state $|\Psi_0\rangle$ violates symmetry under gauge transformations. It will become clear in the following discussions that $|\Psi_0\rangle$ is much simpler than $|\Psi_N\rangle$ and, moreover, manifestly shows essential features of superconductivity. It should be kept in mind that $|\Psi_0\rangle$ is not the exact ground state but a variational wave function, so to speak, in which the parameters u_k and v_k are fixed such that the expectation value $E = \langle\Psi_0|H|\Psi_0\rangle$ of the Hamiltonian (2.17) takes an extremum.

2.3 The mean-field theory of pair condensation

Although the BCS theory has started with the ground state $|\Psi_0\rangle$, it is a mean-field theory based on the assumption that the expectation value of the pair is finite and hence is most clearly explained from that point of view. Here we discuss, for later convenience, the most general pairing state at finite temperatures. Thus, instead of the above expectation values, the statistical averages

$$\begin{aligned}\Psi_{\alpha\beta}(\boldsymbol{k}) &\equiv \langle a_{k\alpha}a_{-k\beta}\rangle \\ \Psi^*_{\alpha\beta}(\boldsymbol{k}) &\equiv \langle a^\dagger_{-k\beta}a_{k\alpha}\rangle\end{aligned} \tag{2.26}$$

are assumed finite. These averages are called the *pair amplitudes*.

Fermi statistics imposes a condition, similar to Eq. (2.7),

$$\Psi_{\beta\alpha}(-\boldsymbol{k}) = -\Psi_{\alpha\beta}(\boldsymbol{k}) \tag{2.27}$$

on the pair amplitude. The mean-field approximation amounts to keeping only the interaction terms which induce the transition between as considered in the Hamiltonian (2.17). The $\boldsymbol{q}$-summation may be dropped since all the pairs are assumed to have vanishing centre of mass momentum. In the mean-field approximation, one rewrites the interaction as

$$\begin{aligned}&\frac{1}{2}\sum V_{\alpha\beta,\alpha'\beta'}\left[\Psi^*_{\alpha\beta}(\boldsymbol{k}) + \left(a^\dagger_{-k\beta}a^\dagger_{k\alpha} - \Psi^*_{\alpha\beta}(\boldsymbol{k})\right)\right] \\ &\quad\times\left[\Psi_{\alpha'\beta'}(\boldsymbol{k}') + \left(a_{k'\alpha'}a_{-k'\beta'} - \Psi_{\alpha'\beta'}(\boldsymbol{k}')\right)\right]\end{aligned}$$

and drops the second term in the square brackets assuming that the fluctuations of aa and $a^\dagger a^\dagger$ from the mean values are small. Let us introduce the mean fields

$$\begin{aligned}\Delta_{\alpha\beta}(\boldsymbol{k}) &\equiv \sum V_{\alpha\beta,\alpha'\beta'}(\boldsymbol{k},\boldsymbol{k}')\Psi_{\alpha'\beta'}(\boldsymbol{k}') \\ \Delta^\dagger_{\beta\alpha}(\boldsymbol{k}) &\equiv \sum V_{\alpha'\beta',\alpha\beta}(\boldsymbol{k}',\boldsymbol{k})\Psi^\dagger_{\alpha'\beta'}(\boldsymbol{k}')\end{aligned} \tag{2.28}$$

associated with the pair amplitudes (2.26). It follows from Eq. (2.2) that the mean field Δ satisfies relations similar to Eq. (2.27),

$$\Delta_{\alpha\beta}(\boldsymbol{k}) = -\Delta_{\beta\alpha}(-\boldsymbol{k}). \tag{2.29}$$

Now the mean-field Hamiltonian $\mathscr{H}_{\rm mf}$ is given by

$$\mathscr{H}_{\rm mf} = \sum_{k,\alpha} \xi_{k\alpha} a^\dagger_{k\alpha} a_{k\alpha} + \frac{1}{2} \sum \left\{ \Delta^\dagger_{\beta\alpha}(\boldsymbol{k}) a_{k\alpha} a_{-k\beta} \right.$$
$$\left. + a^\dagger_{-k\beta} a^\dagger_{k\alpha} \Delta_{\alpha\beta}(\boldsymbol{k}) - \Delta^\dagger_{\beta\alpha}(\boldsymbol{k}) \Psi_{\alpha\beta}(\boldsymbol{k}) \right\}. \tag{2.30}$$

The mean fields create or annihilate particle pairs. In the mean-field theory, the statistical averages such as (2.26) are obtained in terms of $\mathscr{H}_{\rm mf}$ as[1]

$$\langle \cdots \rangle = Z^{-1} {\rm Tr} \left({\rm e}^{-\beta \mathscr{H}_{\rm mf}} \cdots \right), \quad Z \equiv {\rm Tr} \left({\rm e}^{-\beta \mathscr{H}_{\rm mf}} \right). \tag{2.31}$$

2.4 Representation of the mean field Δ_k

Before we diagonalise Eq. (2.30), let us look at the representation of the pair amplitudes and the mean field Δ_k. The pair spin state is classified into either a spin-singlet state or a spin-triplet state, as with two-particle bound state considered in Section 2.1.

Let us see how the pair amplitude (2.26) transforms under spin-space rotations. For a spin 1/2 particle, the unitary matrix which rotates the spin state by an infinitesimal angle $\delta\theta$ around an axis specified by a unit vector $\boldsymbol{n}$ is

$$R = 1 + \frac{{\rm i}}{2} \delta\theta \boldsymbol{n} \cdot \boldsymbol{\sigma}. \tag{2.32}$$

Here $\boldsymbol{\sigma} = (\sigma_1, \sigma_2, \sigma_3)$ are the Pauli matrices,

$$\sigma_1 = \begin{pmatrix} 0 & 1 \\ 1 & 0 \end{pmatrix}, \quad \sigma_2 = \begin{pmatrix} 0 & -{\rm i} \\ {\rm i} & 0 \end{pmatrix}, \quad \sigma_3 = \begin{pmatrix} 1 & 0 \\ 0 & -1 \end{pmatrix}. \tag{2.33}$$

The amplitude $\Psi_{\alpha\beta}$ transforms as

$$\Psi'_{\alpha\beta} = R_{\alpha\gamma} R_{\beta\eta} \Psi_{\gamma\eta} = R_{\alpha\gamma} \Psi_{\gamma\eta} R^T_{\eta\beta}$$

under this rotation. Since $-\sigma_2 \boldsymbol{\sigma} \sigma_2 = \boldsymbol{\sigma}^T$, one finds that

$$\Psi' = \Psi + \frac{{\rm i}}{2} \delta\theta \boldsymbol{n} \cdot (\boldsymbol{\sigma}\Psi - \Psi \sigma_2 \boldsymbol{\sigma} \sigma_2)$$
$$= \Psi + \frac{{\rm i}}{2} \delta\theta \boldsymbol{n} \cdot (\boldsymbol{\sigma} \Psi \sigma_2 - \Psi \sigma_2 \boldsymbol{\sigma}) \sigma_2. \tag{2.34}$$

[1] The symbols T and $\beta \equiv 1/k_{\rm B}T$ will be used to denote the absolute temperature and its inverse, respectively.

Accordingly, one finds a spin-singlet representation $\Psi^{(1)}$, which is invariant under spin-space rotations, and a spin-triplet representation $\Psi^{(3)}$, which transforms as a vector:

$$\Psi^{(1)} = A(\boldsymbol{k})\mathrm{i}\sigma_2 \tag{2.35}$$

$$\Psi^{(3)} = \boldsymbol{A}(\boldsymbol{k}) \cdot \mathrm{i}\boldsymbol{\sigma}\sigma_2. \tag{2.36}$$

If the latter equation is substituted into Eq. (2.34), one obtains $\Psi' = (\boldsymbol{A} + \delta\theta\boldsymbol{n} \times \boldsymbol{A})\cdot(\mathrm{i}\boldsymbol{\sigma}\sigma_2)$ and hence finds that $\boldsymbol{A}$ transforms as a vector under rotations. The $\boldsymbol{k}$-dependence of A ($\boldsymbol{A}$) represents the orbital state of the singlet (triplet) pair. From antisymmetry (2.27), these quantities must satisfy

$$\begin{array}{ll} \text{(singlet state)} & A(-\boldsymbol{k}) = A(\boldsymbol{k}) \\ \text{(triplet state)} & \boldsymbol{A}(-\boldsymbol{k}) = -\boldsymbol{A}(\boldsymbol{k}). \end{array} \tag{2.37}$$

Therefore the most general form of the pair amplitude takes the form

$$\Psi = \Psi^{(1)} + \Psi^{(3)} = \begin{pmatrix} -A_1 + \mathrm{i}A_2 & A_3 + A \\ A_3 - A & A_1 + \mathrm{i}A_2 \end{pmatrix}. \tag{2.38}$$

It is also possible to obtain similar representations for the mean field $\Delta(\boldsymbol{k})$. Provided that the interaction is of the form (2.4), the singlet mean field is

$$\hat{\Delta}(\boldsymbol{k}) = \Delta(\boldsymbol{k})\mathrm{i}\sigma_2$$
$$\Delta(\boldsymbol{k}) = \sum_{k'} V^{(\mathrm{e})}(\boldsymbol{k}, \boldsymbol{k}')A(\boldsymbol{k}') \tag{2.39}$$

while the triplet mean field is

$$\hat{\Delta}(\boldsymbol{k}) = \boldsymbol{\Delta}(\boldsymbol{k}) \cdot \mathrm{i}\boldsymbol{\sigma}\sigma_2$$
$$\boldsymbol{\Delta}(\boldsymbol{k}) = \sum_{k'} V^{(\mathrm{o})}(\boldsymbol{k}, \boldsymbol{k}')\boldsymbol{A}(\boldsymbol{k}'). \tag{2.40}$$

The vector $\boldsymbol{\Delta}(\boldsymbol{k})$ is also called a $\boldsymbol{d}$ vector and, like the vector $\boldsymbol{A}$, transforms as a vector under spin-space rotations. It can be shown easily that

$$\hat{\Delta}^{\dagger}\hat{\Delta} = \boldsymbol{\Delta}^* \cdot \boldsymbol{\Delta} + \mathrm{i}(\boldsymbol{\Delta}^* \times \boldsymbol{\Delta}) \cdot \boldsymbol{\sigma}. \tag{2.41}$$

Accordingly, if the spin quantisation axis is rotated so that $\boldsymbol{\Delta}^* \times \boldsymbol{\Delta} \parallel \hat{z}$, the above equation is diagonalised and the eigenvalues are easily found to be $\boldsymbol{\Delta}^* \cdot \boldsymbol{\Delta} \pm |\boldsymbol{\Delta}^* \times \boldsymbol{\Delta}|$. A pair with $\boldsymbol{\Delta}^* \times \boldsymbol{\Delta} = 0$ ($\neq 0$) is called a unitary (non-unitary) pair.

The fields $\Delta(\boldsymbol{k})$ and $\boldsymbol{\Delta}(\boldsymbol{k})$ are complex numbers in general and their $\boldsymbol{k}$-dependence represents the orbital state of the pair. The orbital state is classified according to the partial wave analysis introduced in Section 2.1. Examples will be analysed in detail in the following chapters.

The kind of pairs which appear in a given system depends on which state minimises the free energy of the system. Coexistence of the singlet pairing and the triplet pairing is not usually considered since in general this raises the free energy. There is, however, no *a priori* reason to exclude this possibility.

2.5 Bogoliubov transformation

The mean-field Hamiltonian $\mathscr{H}_{\text{mf}}$ of the previous section is quadratic in a and $a^\dagger$ and hence may be diagonalised by the *Bogoliubov transformation*

$$a_{k\alpha} = \sum_\beta \left\{ u_{k\alpha\beta}\gamma_{k\beta} + v_{k\alpha\beta}\gamma^\dagger_{-k\beta} \right\}. \tag{2.42}$$

The new operators γ and $\gamma^\dagger$ are required to satisfy the anticommutation relations for fermions. This condition is satisfied if one takes

$$\begin{gathered} u_k^T = u_{-k}, \quad v_k^T = -v_{-k} \\ u_k u_k^\dagger + v_k v_k^\dagger = \hat{1}, \quad u_k v_k - v_k u_k = 0. \end{gathered} \tag{2.43}$$

Here u_k and v_k are 2×2 matrices whose components are $u_{k\alpha\beta}$ and $v_{k\alpha\beta}$, respectively. One may always impose the Hermiticity condition

$$u_k^\dagger = u_k \tag{2.44}$$

on the matrix u without loss of generality in the cases under consideration.

If one writes $\mathscr{H}_{\text{mf}}$ in terms of γ and $\gamma^\dagger$ defined by the Bogoliubov transformation

$$\begin{aligned} a_k &= u_k\gamma_k + v_k\gamma^\dagger_{-k} \\ a_{-k} &= u_k^T\gamma_{-k} - v_k^T\gamma_k^\dagger \end{aligned} \tag{2.42$'$}$$

that satisfies Eqs. (2.43) and (2.44), and requires that terms $\gamma\gamma$ and $\gamma^\dagger\gamma^\dagger$ do not appear in $\mathscr{H}_{\text{mf}}$, one obtains the condition

$$u_k\hat{\xi}_k v_k + v_k\hat{\xi}_{-k}u_k - u_k\hat{\Delta}_k u_k + v_k\hat{\Delta}_k^\dagger v_k = 0. \tag{2.45}$$

Here $\hat{\Delta}_k$ denotes the matrix $\{\Delta_{\alpha\beta}(\boldsymbol{k})\}$ and

$$\hat{\xi}_k = \begin{pmatrix} \xi_{k\uparrow} & 0 \\ 0 & \xi_{k\downarrow} \end{pmatrix}. \tag{2.46}$$

If Eq. (2.45) is satisfied, the Hamiltonian $\mathscr{H}_{\text{mf}}$ becomes

$$\mathscr{H}_{\text{mf}} = \sum_k \gamma_k^\dagger\hat{\varepsilon}_k\gamma_k + \mathscr{H}_{\text{c}}, \tag{2.47}$$

where

$$\hat{\varepsilon}_k \equiv u_k \hat{\xi}_k u_k - v_k \hat{\xi}_{-k} v_k^\dagger + u_k \hat{\Delta}_k v_k^\dagger + v_k \hat{\Delta}_k^\dagger u_k \tag{2.48}$$

and

$$\mathscr{H}_\mathrm{c} = -\frac{1}{2} \sum_k \mathrm{tr} \left\{ \hat{\varepsilon}_k - u_k \hat{\xi}_k u_k - v_k \hat{\xi}_k v_k^\dagger - \hat{\Delta}_k^\dagger \hat{\Psi}_k \right\} \tag{2.49}$$

denotes the pair condensation energy. (In the above equation, tr denotes the trace over the spin indices.) The matrix $\hat{\varepsilon}_k$ is Hermitian and hence can be considered as a diagonal matrix. Accordingly, it describes the excitation energy of the quasi-particles described by γ and $\gamma^\dagger$. The ground state of the Hamiltonian is the Fock vacuum of these quasi-particles. Henceforward, the term quasi-particles implies these excitations in a superconducting state, unless otherwise stated.

Let us write down the solution of the above problem. General pairings being rather complicated, let us assume that the normal state electron energy is independent of the spin ($\xi_{k\sigma} = \xi_k$) and $\hat{\Delta}$ is unitary, that is, either the pair is singlet or $\hat{\Delta}_k^\dagger \hat{\Delta}_k = \mathbf{\Delta}_\mathbf{k}^* \mathbf{\Delta}_\mathbf{k} \cdot \hat{\mathbf{1}}$. If one employs an *Ansatz*

$$u_{k\alpha\beta} = \bar{u}_k \delta_{\alpha\beta} / D_k, \quad v_{k\alpha\beta} = +\Delta_{k\alpha\beta} / D_k$$

the condition (2.45) becomes $2\xi_k \bar{u}_k - \bar{u}_k^2 + |\Delta_k|^2 = 0$, where $|\Delta_k|^2 \equiv \frac{1}{2}\mathrm{tr}\left(\hat{\Delta}_k \hat{\Delta}_k^\dagger\right)$. From this and Eq. (2.48) one obtains

$$\begin{aligned} \bar{u}_k &= \xi_k + \varepsilon_k \\ D_k^2 &= (\xi_k + \varepsilon_k)^2 + |\Delta_k|^2 \\ \varepsilon_k &= \sqrt{\xi_k^2 + |\Delta_k|^2}. \end{aligned} \tag{2.50}$$

After some algebra, one finally obtains the Bogoliubov transformation matrices

$$u_k = \frac{1}{\sqrt{2}} \sqrt{1 + \frac{\xi_k}{\varepsilon_k}} \cdot \hat{1}, \quad v_k = \frac{1}{\sqrt{2}} \sqrt{1 - \frac{\xi_k}{\varepsilon_k}} \cdot \frac{\hat{\Delta}_k}{|\Delta_k|}. \tag{2.51}$$

Therefore, $\mathscr{H}_\mathrm{mf}$ is diagonalised by a transformation of the above form for a spin-singlet pairing as well as a spin-triplet pairing provided it is unitary. (See [E-9] for diagonalisation in general cases.)

The final step in the mean-field approximation is to find the pair amplitude Ψ_k and hence the mean field Δ_k. The statistical average (2.26) for the pair amplitude is easily obtained from $\mathscr{H}_\mathrm{mf}$ of the form (2.47) and the fact that γ and $\gamma^\dagger$ satisfy the Fermionic anti-commutation relations. If the result is substituted into Eq. (2.28), one obtains

$$\Delta_{k\alpha\beta} = -\sum_{k',\alpha'\beta'\gamma'} V_{\alpha\beta,\alpha'\beta'}(\boldsymbol{k},\boldsymbol{k}') \left\{ u_{k'\alpha'\gamma'} v_{k'\gamma'\beta'} \left[1 - f(\varepsilon_{k'\gamma'})\right] \right.$$
$$\left. - v_{k'\alpha'\gamma'} u_{k'\gamma'\beta'} f(\varepsilon_{-k'\gamma'}) \right\}. \tag{2.52}$$

Here

$$f(x) = \left(\mathrm{e}^{\beta x} + 1\right)^{-1} = \frac{1}{2}\left(1 - \tanh\frac{\beta x}{2}\right) \tag{2.53}$$

is the Fermi distribution function. Equation (2.52) gives the mean field Δ_k in a self-consistent manner and is often called the *gap equation.*

Before we close this section, let us write down the free energy. This is given by

$$F(T,V,\mu) = -\beta^{-1}\ln\mathrm{Tr}\left(\mathrm{e}^{-\beta\mathscr{H}}\right) = -\beta^{-1}\sum_{k,\alpha}\ln\left(1+\mathrm{e}^{-\beta\varepsilon_{k\alpha}}\right) + \mathscr{H}_{\mathrm{c}}. \tag{2.54}$$

The symbol Tr stands for the trace over all possible states of the system. If Eqs. (2.50) and (2.51) are applicable, Eq. (2.54) becomes

$$F_0 = -\beta^{-1}\sum_{k,\alpha}\ln\left(1+\mathrm{e}^{-\beta\varepsilon_{k\alpha}}\right) - \sum_{k,\alpha}\left(\varepsilon_{k\alpha} - \xi_{k\alpha} + \frac{1}{2}\mathrm{tr}\hat{\Delta}_k^{\dagger}\hat{\Psi}_k\right). \tag{2.55}$$

Let us note that the gap equation (2.52) is equivalent to the condition that $\hat{\Delta}_k$ minimises the free energy, namely

$$\frac{\partial F}{\partial \hat{\Delta}_k^*} = 0. \tag{2.56}$$

The matrix elements of Δ_k and $\Delta_k^{\dagger}$ are assumed to be independent when the above variation is taken. The free energy F may be rewritten as

$$\begin{aligned} F &= E - TS \\ S &= -k_B\sum_{k,\alpha}\{(1-f)\ln(1-f) + f\ln f\} \\ E &= \sum_{k,\alpha}\varepsilon_{k\alpha} f + \mathscr{H}_{\mathrm{c}}, \end{aligned} \tag{2.57}$$

where $f = f(\varepsilon_{k\alpha})$. S is the entropy of thermally excited quasi-particles and E is the internal energy, the first term of which is the quasi-particle excitation energy while the second term is the condensation energy.

When the interaction is of the form (2.4) and hence conserves the total spin, Eq. (2.52) (for unitary states) takes the same form,

$$\Delta_{k\alpha\beta} = -\sum_{k'} V(\boldsymbol{k},\boldsymbol{k}')\frac{\Delta_{k'\alpha\beta}}{2\varepsilon_{k'\alpha}}\left[1 - 2f(\varepsilon_{k'\alpha})\right] \tag{2.58}$$

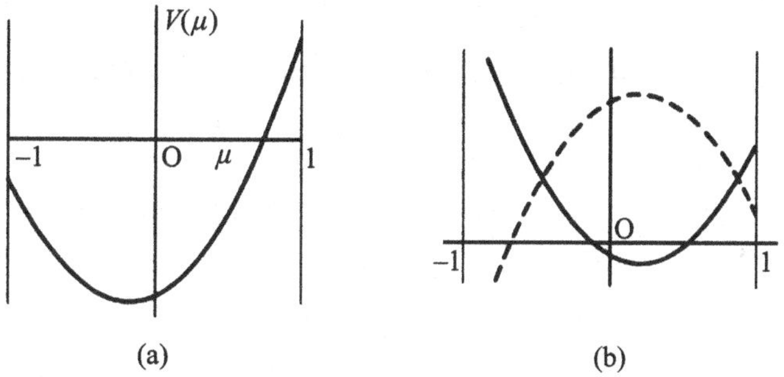

Fig. 2.2. (a) An example of $V(\mu)$ which leads to s-wave pairing. (b) An example of $V(\mu)$ which leads to p-wave (solid line) or d-wave (broken line) pairing.

for both spin-singlet pairing and spin-triplet pairing. This can be easily shown when Eqs. (2.50) and (2.51) are applicable.

The gap $\Delta_{k\alpha\beta}$ takes a nonvanishing value at the critical temperature T_c. T_c is obtained by putting $\varepsilon_k = |\xi_k|$ in Eq. (2.58). When the interaction is of the form (2.12) and, furthermore, V_l is regarded as constant in the vicinity of the Fermi surface, that is

$$V_l = \begin{cases} \text{constant} & (|\xi_k|, |\xi_{k'}| \le \omega_c) \\ 0 & (|\xi_k|, |\xi_{k'}| > \omega_c), \end{cases} \tag{2.59}$$

then the integration over the direction of $\boldsymbol{k}'$ may be carried out. Finally one obtains the following equation for each l, which determines the critical temperature $T_c = 1/k_B\beta_c$,

$$1 = -N(0)V_l \int_0^{\omega_c} d\xi \frac{1}{\xi} \tanh \frac{\beta_c \xi}{2}. \tag{2.60}$$

Therefore, in this simplified model, the equation which determines T_c takes the same form irrespective of the type of pair formation. Therefore it is expected that condensation takes place in the l-channel which has the largest attractive $|V_l|$ and hence has the highest T_c. Since the gap equation (2.58) is nonlinear in $\Delta_{k\alpha\beta}$ at $T < T_c$, in general, there may appear several components with different l. However, in practice, the proportion of components other than the dominant one is small.

To find the values of V_l and, in particular, which V_l is negative and most attractive, one has to know the effective interaction between particles. This is generally a difficult problem. Just as an example, suppose the interaction V takes the form

$$V(\mu) = V_0 + 3V_1P_1(\mu) + 5V_2P_2(\mu)$$

on the Fermi surface. Here $|\boldsymbol{k}| \sim |\boldsymbol{k}'| \sim k_F$ and $\mu = \hat{\boldsymbol{k}} \cdot \hat{\boldsymbol{k}}'$. If the interaction is attributed to phonon exchanges or spin fluctuation exchanges, its wave number is $\boldsymbol{q} = \boldsymbol{k}' - \boldsymbol{k}$ and μ becomes $\mu = \hat{\boldsymbol{k}} \cdot \hat{\boldsymbol{k}}' = 1 - \left(q^2/2k_F^2\right)$, see Fig. 2.1. Figure 2.2 shows $V(\mu)$ for the cases (a) $V_1 = V_2 = 1$ and (b) $V_1 = -1, V_2 = 1$ (solid line) and $V_1 = 1, V_2 = -1$ (broken line), with an arbitrary V_0. They lead to condensation with s-wave, p-wave and d-wave pairs, respectively.

3
BCS theory

Basic properties of a superconducting state according to the so-called BCS theory will be discussed in the present chapter. According to this theory, an attractive force between electrons in the vicinity of the Fermi surface leads to superconductivity with spin-singlet s-wave pairing. Our discussion here serves as a basis for the generalisations presented in the following chapters.

3.1 Spin-singlet pairing and the energy gap

The interaction in an ordinary superconductor is thought to be attractive down to the average distance between conduction electrons, see Chapter 4. If this is the case, the amplitude independent of the angle, the s-wave (l=0) amplitude, is the most important in Eq. (2.12). Therefore, as mentioned in Section 2.2, ordinary superconductivity is caused by the spin-singlet s-wave pairings. In this case, the mean field Δ_k takes the form (2.39), namely

$$\hat{\Delta}_k = \begin{pmatrix} 0 & \Delta_k \\ -\Delta_k & 0 \end{pmatrix} \tag{3.1}$$

and $\Delta_k = \Delta_{-k}$ is in general a complex number.

The Bogoliubov transformation

$$\begin{aligned} a_{k\uparrow} &= u_k\gamma_{k\uparrow} + v_k\gamma^{\dagger}_{-k\downarrow} \\ a_{-k\downarrow} &= u_k\gamma_{-k\downarrow} - v_k\gamma^{\dagger}_{k\uparrow} \end{aligned} \tag{3.2}$$

for the above $\hat{\Delta}_k$ is already given in Eq. (2.51) as

$$u_k = \frac{1}{\sqrt{2}}\left(1+\frac{\xi_k}{\varepsilon_k}\right)^{1/2}\hat{1}, \quad v_k = \frac{1}{\sqrt{2}}\left(1-\frac{\xi_k}{\varepsilon_k}\right)^{1/2}\frac{\hat{\Delta}_k}{|\Delta_k|}. \tag{3.3}$$

As was mentioned in Section 2.2, the BCS theory assumes that there is an attractive s-wave interaction of constant amplitude between electrons near the Fermi surface. That is, it is assumed, as in Eq. (2.59), that

$$V_0(k, k') = \begin{cases} -g & (\omega_c > |\xi_k|, |\xi_{k'}|) \\ 0 & \text{(otherwise)} \end{cases} \tag{3.4}$$

where g is a positive constant and ω_c is of the order of the Debye frequency if the attractive force is mediated by phonons. For the above interaction, the gap takes a constant value $\Delta_k = \Delta$ for $|\xi_k| < \omega_c$ and vanishes otherwise. The constant Δ is determined by the gap equation (2.58), which takes the form

$$\Delta = g \sum_{|\xi_{k'}|<\omega_c} \frac{\Delta}{2\varepsilon_{k'}} \tanh \frac{\beta \varepsilon_{k'}}{2}. \tag{3.5}$$

Here ε_k is the excitation energy of quasi-particles,

$$\varepsilon_k = \sqrt{\xi_k^2 + |\Delta|^2}. \tag{3.6}$$

Let us consider an excitation on the ground state, that is, the quasi-particle vacuum, to simplify our arguments. The inverse of the Bogoliubov transformation tells us that we need to add an electron to the system or subtract an electron from the system to excite a single quasi-particle. Accordingly, one has to excite a quasi-particle pair if the process is to keep the electron number of the system fixed. Therefore, at least $2|\Delta_0|$ of energy is necessary for the excitation to take place, Δ_0 being Δ at $T = 0$. In other words, the excitation spectrum has a gap. A pair must be broken to create an excitation in the ground state, where all the electrons are paired to form the condensate. Hence at least the binding energy of the pair is required. At finite temperatures, there already exists a gas of thermally excited quasi-particles and the scattering process of the quasi-particles must also be considered. Quasi-particle creation in this case is the same as that for $T = 0$, except that there are particles which do not form pairs and hence the mean field is suppressed and the gap $\Delta(T)$ becomes less than Δ_0.

A quasi-particle is specified by the momentum $\boldsymbol{k}$ and the spin α. The density of states of quasi-particles with spin α in the vicinity of the Fermi surface is given by

$$\begin{aligned} \mathscr{D}(\omega) &= \sum_k \delta(\varepsilon_k - |\omega|) \simeq N(0) \int \mathrm{d}\xi_k \delta(\varepsilon_k - |\omega|) \\ &= N(0) \frac{|\omega|}{\sqrt{\omega^2 - |\Delta|^2}} \theta(|\omega| - |\Delta|). \end{aligned} \tag{3.7}$$

According to the above equation, the states within the gap are pushed outside

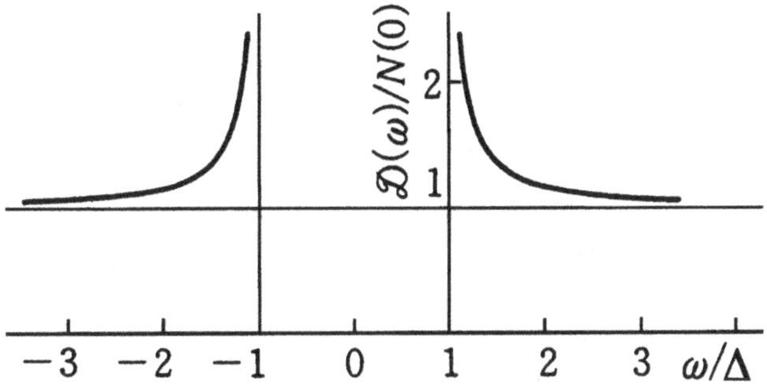

Fig. 3.1. The density of states $\mathscr{D}(\omega)$ in a superconducting state.

the gap as in Fig. 3.1 and the density of states diverges as $1/\sqrt{\omega^2 - |\Delta|^2}$ as $\omega \to \pm|\Delta|$. It will be shown later that the density of states appears in various physical properties of superconductors. It should be noted that Eq. (3.7) is correct only for an isotropic Fermi surface and that $\mathscr{D}(\omega)$ deviates from this equation in an actual metal since the modulus of Δ varies on the Fermi surface. Note also that the group velocity of the quasi-particle excitation is $v_g(\boldsymbol{k}) = \nabla_k \varepsilon_k = (\boldsymbol{k}/m)(\xi_k/\varepsilon_k)$.

3.1.1 Energy gap

The factor Δ on both sides of Eq. (3.5) has been retained to show that there always exists a solution $\Delta = 0$, but it will be dropped hereafter. Since $\omega_c \ll \varepsilon_F$, Eq. (3.5) becomes

$$1 = N(0)g \int_0^{\omega_c} d\xi \frac{1}{\varepsilon} \tanh \frac{\beta\varepsilon}{2}. \qquad (3.8)$$

Let us first consider the limit $\beta = \infty$, that is, $T = 0$ and hence find the gap Δ_0 in the ground state. (The absolute value symbol for Δ will be dropped hereafter.) When $\omega_c \gg \Delta_0$, it takes a form identical to Eq. (2.21), that is,

$$\Delta_0 = 2\omega_c \exp[-1/N(0)g]. \qquad (3.9)$$

Next let us consider the critical temperature T_c at which the nonvanishing solution Δ appears for the first time as the system is cooled down. For calculational convenience, Eq. (3.8) is rewritten using the formula

$$\frac{1}{2x} \tanh \frac{x}{2} = \sum_{n=-\infty}^{\infty} \frac{1}{(2n+1)^2\pi^2 + x^2} \qquad (3.10)$$

as

$$1 = 2N(0)g\beta^{-1}\sum_{n=-\infty}^{\infty}\frac{1}{\sqrt{\omega_n^2+|\Delta|^2}}\tan^{-1}\frac{\omega_c}{\sqrt{\omega_n^2+|\Delta|^2}}, \tag{3.11}$$

where $\omega_n = (2n+1)\pi\beta^{-1}$. Since the gap Δ vanishes at $T = T_c = 1/\beta_c$, one has an equation that determines β_c:

$$1 = 4N(0)g\sum_{n=0}^{\infty}\frac{1}{(2n+1)\pi}\tan^{-1}\frac{\beta_c\omega_c}{(2n+1)\pi}.$$

Since $\beta_c\omega_c \gg 1$, the summation in the right hand side may be approximated by $\frac{1}{2}\sum_{n\le n_c}(2n+1)^{-1}$, where $(2n_c+1)\pi = \beta_c\omega_c$. Then one obtains

$$\begin{aligned}[N(0)g]^{-1} &= \psi(\beta_c\omega_c/2\pi+1) - \psi(1/2)\\ &\simeq \ln(\beta_c\omega_c/2\pi) - \ln(e^{-\gamma}/4),\end{aligned} \tag{3.12}$$

where $\psi(z)$ is the di Gamma function and use has been made of the formula

$$\sum_{m=1}^{n}\frac{1}{m+z} = \psi(n+z+1) - \psi(z+1),$$

the asymptotic form

$$\psi(z) \simeq \ln z + O(|z|^{-1}),$$

and the value

$$\psi(1/2) = \ln(e^{-\gamma}/4).$$

From these expressions one finally obtains the BCS expression for the critical temperature

$$k_B T_c = 1.13\omega_c \exp\left[-1/N(0)g\right], \tag{3.13}$$

where the numerical value $2e^{\gamma}/\pi \simeq 1.13$ has been used. From this expression and Eq. (3.9) one obtains the ratio

$$\frac{2\Delta_0}{k_B T_c} = 3.53. \tag{3.14}$$

It is expected physically that $k_B T_c$ is of the order of the pair-breaking energy. The ratio 3.53 is a 'universal' value independent of T_c and other variables. It should be noted, however, that this value is obtained within the so-called weak coupling approximation in the BCS theory. As is shown in Table 3.1, the actual value for ordinary superconducting metals with low T_c is close to this weak coupling value but for metals with higher T_c, this ratio clearly deviates from 3.53. This deviation from the BCS value will be an important point in the following chapters.

Table 3.1. *Observed values of* $2\Delta_0/k_B T_c$ *and* $\Delta C/C_n$

	T_c	$2\Delta_0/k_B T_c$	$\Delta C/C_n$
BCS		3.53	1.43
Al	1.18	3.53	1.45
Cd	0.52	3.44	1.32
Sn	3.72	3.61	1.60
Hg	4.15	3.65	2.37
Pb	7.20	3.95	2.71
Nb	9.25	3.65	1.87

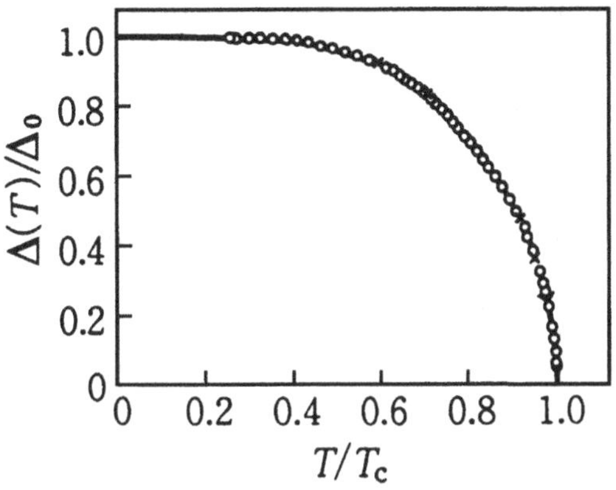

Fig. 3.2. The temperature dependence of the energy gap of Pb. The symbol ∘ denotes values obtained by the tunnelling experiment. The solid line is the prediction of the BCS theory.

Figure 3.2 shows the temperature dependence of the energy gap, from which one finds that the mean-field theory reproduces the data very well. At low temperatures $T \ll T_c$ the gap is given approximately as

$$\Delta(T) \simeq \Delta_0 \left[1 - \left(\frac{2\pi}{\beta\Delta_0}\right)^{1/2} e^{-\beta\Delta_0}\right]. \tag{3.15}$$

The behaviour of $\Delta(T)$ in the vicinity of T_c will be given in the next section.

3.1.2 Coherence length

The energy scale which characterises a superconducting state is Δ. The corresponding characteristic length

$$\xi(T) \equiv \hbar v_F/\pi\Delta(T), \tag{3.16}$$

called the *coherence length*, is important in the analysis of superconducting phenomena. Let us write the coherence length at $T = 0$ as ξ_0. The pair wave function (2.26) of the condensate is $\Psi_k = -u_k v_k$, which is given, in real space, by

$$\Psi(x) = \sum_k \Psi_k \exp(ik \cdot x) = \sum_k \frac{1}{2}\left(\frac{|\Delta|^2}{\xi_k^2 + |\Delta|^2}\right)^{1/2} \exp(ik \cdot x).$$

This is a similar form to the Cooper pair wave function considered in Section 2.2 and shows that the coherence length ξ represents the spatial extension of a pair. The length ξ_0 is of the order of 10^{-4} cm in an ordinary superconducting metal, which is much larger than the distance between electrons. Accordingly many pairs overlap one another.

3.1.3 Charge density

The charge density en is obtained with the help of the Bogoliubov transformation (3.2) as

$$\begin{aligned} e\sum_{k,\alpha}\langle a_{k\alpha}^{\dagger}a_{k\alpha}\rangle &= Q_s + Q^*, \\ Q_s = 2\sum_k |v_k|^2, \quad Q^* &= 2\sum_k \left(|u_k|^2 - |v_k|^2\right) f(\varepsilon_k). \end{aligned} \tag{3.17}$$

The first term Q_s may be thought of as the contribution of the condensate i.e., the pairs and is given by the number of states $n = k_F^3/3\pi^2$ with $\xi_k < 0$. The second term Q^* is the contribution of the quasi-particle excitations and vanishes for an equilibrium Fermi distribution with particle–hole symmetry. If, on the other hand, the system is out of equilibrium and particle excitations (above the Fermi level) and hole excitations (below the Fermi level) obey different distributions, the second term fails to vanish. One may consider, in this case, that a quasi-particle with a momentum $\boldsymbol{k}$ has a charge $q_k = e\left(|u_k|^2 - |v_k|^2\right) = e\xi_k/\varepsilon_k$. It should be noted that by calculating $\partial \sum_k |v_k|^2/\partial\mu$, one finds that Q_s changes by $\delta Q_s = 2eN(0)\delta\mu$ under an infinitesimal change $\delta\mu$ of the chemical potential.

3.2 Thermodynamic properties

The thermodynamic properties of a superconductor are derived from the free energy (2.55). Before we do this, however, let us consider the thermodynamic relations relevant to a superconductor in a perfectly diamagnetic state. Consider a long cylinder made of a superconductor and suppose that a uniform external magnetic field of strength H is applied parallel to the axis of the cylinder. The magnetic flux density B is also uniform and parallel to the axis of the cylinder. Let us denote the superconducting state and the normal state by the indices s and n respectively, and the free energy and entropy per unit volume by F and S respectively. One has $\mathrm{d}F = -S\mathrm{d}T + (1/4\pi)\boldsymbol{H}\cdot\mathrm{d}\boldsymbol{B}$ since the work required to change the flux density from $\boldsymbol{B}$ to $\boldsymbol{B}+\mathrm{d}\boldsymbol{B}$ is $(1/4\pi)\boldsymbol{H}\cdot\mathrm{d}\boldsymbol{B}$ per unit volume. If $\boldsymbol{H}$ is taken to be an independent variable, one should use the free energy $G(T,\boldsymbol{H}) = F(T,\boldsymbol{B}) - \boldsymbol{B}\cdot\boldsymbol{H}/4\pi$, instead of $F(T,\boldsymbol{B})$ (the other variables are omitted). One therefore obtains

$$G(T,\boldsymbol{H}) - G(T,0) = -\frac{1}{4\pi}\int_0^{H} \boldsymbol{B}(\boldsymbol{H}')\cdot\mathrm{d}\boldsymbol{H}'. \tag{3.18}$$

In the normal state, one may put $\boldsymbol{B} = \boldsymbol{H}$ since the magnetic susceptibility of an ordinary metal is of the order of $10^{-7} \sim 10^{-6}$ emu. One may also take $\boldsymbol{B} = 0$ in the superconducting state if perfect diamagnetism is assumed. Then one has

$$\begin{aligned} G_{\mathrm{n}}(T,\boldsymbol{H}) &= G_{\mathrm{n}}(T,0) - \frac{1}{8\pi}\boldsymbol{H}^2 \\ G_{\mathrm{s}}(T,\boldsymbol{H}) &= G_{\mathrm{s}}(T,0) = F_{\mathrm{s}}(T). \end{aligned} \tag{3.19}$$

Let us focus on type 1 superconductors here. (Type 2 superconductors will be studied in Chapter 5.) Since a superconductor becomes normal when the external magnetic field reaches $H = H_{\mathrm{c}}(T)$, one has $G_{\mathrm{s}} = G_{\mathrm{n}}$ at that field and hence one finds $G_{\mathrm{s}}(T,0) - G_{\mathrm{n}}(T,0) = -H_{\mathrm{c}}^2(T)/8\pi$. This therefore implies that the free energy difference at $\boldsymbol{H} = 0$ is found by measurement of the critical field. The free energy difference under a magnetic field $\boldsymbol{H}$ is obtained from the above relation and Eq. (3.19) as

$$G_{\mathrm{s}}(T,\boldsymbol{H}) - G_{\mathrm{n}}(T,\boldsymbol{H}) = \frac{1}{8\pi}(\boldsymbol{H}^2 - H_{\mathrm{c}}^2(T)), \tag{3.20}$$

from which the entropy difference is found to be

$$S_{\mathrm{s}}(T,\boldsymbol{H}) - S_{\mathrm{n}}(T,\boldsymbol{H}) = \frac{1}{4\pi}H_{\mathrm{c}}(T)\frac{\mathrm{d}H_{\mathrm{c}}(T)}{\mathrm{d}T}.$$

This is negative as it should be. One finds the specific heat jump at T_{c} by differentiating the above equation with respect to T and using the identity $H_{\mathrm{c}}(T_{\mathrm{c}}) = 0$,

$$\Delta C = (C_{\mathrm{s}} - C_{\mathrm{n}})_{T=T_{\mathrm{c}}} = \frac{T_{\mathrm{c}}}{4\pi}\left[\left(\frac{\mathrm{d}H_{\mathrm{c}}}{\mathrm{d}T}\right)_{T_{\mathrm{c}}}\right]^2. \tag{3.21}$$

3.2.1 The free energy according to the BCS theory

Let us consider the free energy of the BCS model of the previous section. One has to calculate the difference of the free energy of the superconducting state ($\Delta \neq 0$) and of the normal state ($\Delta = 0$), supposing that it should exist. We are interested in this difference because the free energy F_s contains the contribution from the interior of the Fermi sphere, for which $|\xi_k|$ is large, and hence the free energy per particle is a large quantity of the order of ε_F. Accordingly the evaluation of F_s would require the quantitative theory of the normal state of a metal. However the difference of F_s and F_n may be discussed within the present model. Equation (2.55) is given in the present case by

$$F_s = -\sum_k \left\{2\beta^{-1} \ln\left(1 + e^{-\beta\varepsilon_k}\right) + \varepsilon_k - \xi_k + \Delta_k^* \Psi_k\right\} \tag{3.22}$$

and one obtains, in the BCS model,

$$\begin{aligned} F_s - F_n &= F_s + 2\sum_{k<k_F} |\xi_k| + \sum_k 2\beta^{-1} \ln\left(1 + e^{-\beta|\xi_k|}\right) \\ &= -4N(0)\beta^{-1} \int_0^\infty d\xi \ln \frac{1 + e^{-\beta\varepsilon}}{1 + e^{-\beta|\xi|}} \\ &\quad -2N(0) \int_0^{\omega_c} d\xi (\varepsilon - |\xi|) + |\Delta|^2/g. \end{aligned} \tag{3.23}$$

The amplitude $|\Delta(T)|$ will be simply written as Δ in the remainder of this section. If Eq. (3.9) is used, one obtains

$$\begin{aligned} F_s - F_n = &-\beta^{-1} 4N(0) \left\{ \int_0^\infty d\xi \ln\left(1 + e^{-\beta\varepsilon}\right) - \frac{\pi^2}{12}\beta^{-1} \right\} \\ &-N(0)\Delta^2 \left(\frac{1}{2} + \ln \frac{\Delta_0}{\Delta}\right). \end{aligned} \tag{3.24}$$

1. $T \ll T_c$: The integral in Eq. (3.24) becomes, upon partial integration,

$$\beta\Delta^2 \int_1^\infty dx \sqrt{x^2 - 1} \left(e^{\beta\Delta x} + 1\right)^{-1},$$

which is the modified Bessel function $K_1(\beta\Delta)$ for $\beta\Delta \gg 1$. If the asymptotic form of K_1 is employed and the second term is evaluated within the lowest order in $1 - (\Delta/\Delta_0)$, Eq. (3.24) becomes

$$\begin{aligned} F_s - F_n(T = 0) \simeq &-\frac{1}{2} N(0)\Delta_0^2 - 2N(0)(2\pi\Delta_0\beta^{-3})^{1/2} e^{-\beta\Delta_0} \\ &+\frac{\pi^2}{3} N(0)(k_B T)^2. \end{aligned} \tag{3.25}$$

The third term is the contribution from the normal state, from which one obtains the electronic specific heat of the normal state,

$$C_{\rm n} = \frac{2}{3}\pi^2 N(0) k_{\rm B}^2 T. \tag{3.26}$$

The coefficient of T on the right hand side is commonly called the Sommerfeld constant and denoted by γ. From the second term of Eq. (3.25), the specific heat of the superconducting state can be found:

$$C_{\rm s} \simeq 2N(0)k_{\rm B}(2\pi\Delta_0^5\beta^3)^{1/2}{\rm e}^{-\beta\Delta_0}. \tag{3.27}$$

This specific heat is exponentially small at low temperatures due to the energy gap. From Eqs. (3.20), (3.25) and (3.14), one obtains

$$\begin{aligned} H_{\rm c}^2(0)/8\pi &= N(0)\Delta_0^2/2 \\ H_{\rm c}(T) &= H_{\rm c}(0)\left[1 - \frac{{\rm e}^{2\gamma}}{3}\left(\frac{T}{T_{\rm c}}\right)^2\right]. \end{aligned} \tag{3.28}$$

It is physically understandable that the condensation energy at $T = 0$ is of the order of the product of the number of pairs $N(0)\Delta_0/2$ and the pair binding energy Δ_0. We note that the temperature dependence of the critical magnetic field (3.28) is valid over quite a wide temperature range.

2. $T \lesssim T_{\rm c}$: Let us next consider the temperatures in the vicinity of $T_{\rm c}$. We will use the results obtained here when we discuss the Ginzburg–Landau theory later. Since $\Delta(T)/k_{\rm B}T_{\rm c} \ll 1$ near $T = T_{\rm c}$, the free energy may be expanded in powers of Δ. Let us recall that Δ_k is a complex number and hence Δ_k and Δ_k^* are independent quantities. For this reason it is easier to compute

$$\frac{\partial F_{\rm s}}{\partial \Delta_k^*} = -\left\{[1 - 2f(\varepsilon_k)]\frac{\partial \varepsilon_k}{\partial \Delta_k^*} - \Psi_k\right\} \tag{3.29}$$

by differentiating Eq. (3.22) rather than by evaluating directly $F_{\rm sn} \equiv F_{\rm s} - F_{\rm n}$. Hence one finds in the BCS model

$$\begin{aligned} \frac{1}{\Delta}\frac{\partial F_{sn}}{\partial \Delta^*} &= -\frac{1}{2}\sum_k \left\{\frac{1}{\varepsilon_k}\tanh\frac{\beta\varepsilon_k}{2} - \frac{1}{\xi_k}\tanh\frac{\beta_{\rm c}\xi_k}{2}\right\} \\ &\simeq 2N(0)\int_0^\infty {\rm d}\xi \frac{1}{\xi}\,[f(\beta\xi) - f(\beta_{\rm c}\xi)] \\ &\quad + \beta^{-1}N(0)|\Delta|^2 \sum_n \int_0^\infty {\rm d}\xi \frac{1}{(\omega_{\rm nc}^2 + \xi^2)^2}, \end{aligned}$$

where $\omega_{nc} = (2n+1)\pi/\beta_c$. Equation (3.11) has been used to obtain the second expression. The first term of the second expression is readily evaluated if one notes that $(1 - T/T_c) \ll 1$. The second term is also obtained easily if ξ-integration is carried out first. Finally one obtains an expression

$$F_{sn} = -a|\Delta|^2 + \frac{b}{2}|\Delta|^4$$
$$a = N(0)(1 - T/T_c), \quad b = \frac{7\zeta(3)}{8\pi^2} N(0)\beta_c^2, \tag{3.30}$$

where $\zeta(3) = \sum_{n=1}^{\infty} n^{-3}$. As was noted at the end of Section 2.4, the equilibrium value of the gap $|\Delta|$ is determined so as to minimise F_{sn} and hence one finds

$$|\Delta|^2 = \frac{a}{b} = \frac{8\pi^2}{7\zeta(3)} (k_B T_c)^2 \left(1 - \frac{T}{T_c}\right). \tag{3.31}$$

By making use of the above result, the equilibrum free energy is found to be

$$F_{sn} = -\frac{1}{8\pi} H_c^2(T) = -\frac{a^2}{2b}$$
$$= -\frac{1}{2} N(0) |\Delta|^2 \left(1 - \frac{T}{T_c}\right)^2. \tag{3.32}$$

It should be noted that F_{sn} is proportional to $(1 - T/T_c)^2$. It then follows from this result that the jump in the specific heat at T_c is given by

$$\Delta C / C_n = 12/7\zeta(3) = 1.43. \tag{3.33}$$

This value is universal in the BCS model. The comparison between the universal number and the measured value is made in Table 3.1. The temperature dependence of C_s in the vicinity of T_c comes from the next order term in the expansion and hence is proportional to $T/T_c - 1$. Figure 3.3 shows the variation in the specific heat of aluminium, and shows an excellent agreement between experiment and BCS theory.

3.2.2 States carrying supercurrent

Pair condensation takes place with the centre of mass momentum $\boldsymbol{q} \neq 0$ in a state with a superconducting current. Therefore the thermodynamic average $\Psi_{\alpha\beta}(\boldsymbol{k}, \boldsymbol{q}) = \langle a_{k_+\alpha} a_{-k_-\beta} \rangle$ becomes finite, where $\boldsymbol{k}_\pm \equiv \boldsymbol{k} \pm \boldsymbol{q}/2$. This

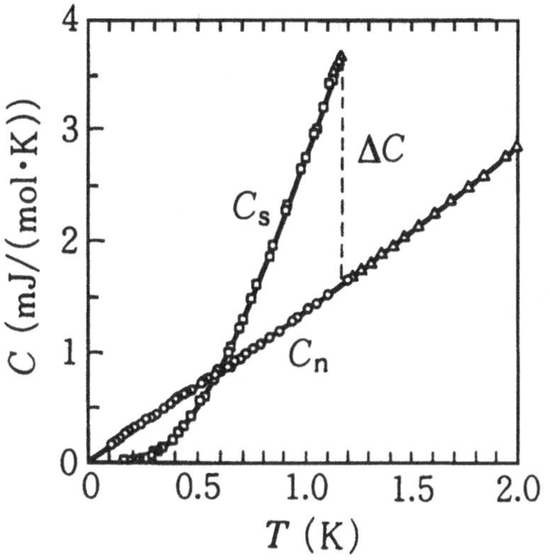

Fig. 3.3. The temperature dependence of the specific heat of aluminium. The specific heat of the normal state is observed to be proportional to T under a magnetic field of 300 G. (From [G-5].)

state is an equilibrium state with the specified total momentum. Therefore it is still a thermodynamic state. The real space order parameter in this case, $\Psi(\boldsymbol{x}) = \sum_k \mathrm{e}^{\mathrm{i}(\boldsymbol{k}_+ - \boldsymbol{k}_-)\cdot\boldsymbol{x}} \langle a_{k_+\uparrow} a_{k_-\downarrow} \rangle$, has a phase factor $\mathrm{e}^{\mathrm{i}\boldsymbol{q}\cdot\boldsymbol{x}}$. Noting that $|q| \ll k_{\mathrm{F}}$, one finds that the electron energies taking part in the pair formation are

$$
\begin{aligned}
\xi_{k_+\uparrow} &= \xi_k + \boldsymbol{k}\cdot\boldsymbol{q}/2m \\
\xi_{k_+\downarrow} &= \xi_k - \boldsymbol{k}\cdot\boldsymbol{q}/2m.
\end{aligned}
\tag{3.34}
$$

Theories developed in Chapter 2 are generalised easily in the present case and one obtains the following results, in which the coefficients of the Bogoliubov transformation

$$
\begin{aligned}
a_{k_+\uparrow} &= u_k \gamma_{k_+\uparrow} + v_k \gamma^\dagger_{-k_-\downarrow} \\
a_{-k_-\downarrow} &= u_k \gamma_{-k_-\downarrow} - v_k \gamma^\dagger_{-k_+\uparrow}
\end{aligned}
\tag{3.35}
$$

are

$$
u_k = \frac{1}{\sqrt{2}}\left(1 + \frac{\xi_k}{\varepsilon_k}\right)^{1/2}, \quad v_k = \frac{1}{\sqrt{2}}\left(1 - \frac{\xi_k}{\varepsilon_k}\right)^{1/2} \frac{\Delta_k}{|\Delta_k|}.
$$

Formally they take the same form as the case with $\boldsymbol{q} = 0$. It should be noted, however, that the excitation energies are modified to

$$\begin{aligned} \varepsilon_{k+\uparrow} &= \varepsilon_k + \boldsymbol{k}\cdot\boldsymbol{q}/2m \\ \varepsilon_{-k-\downarrow} &= \varepsilon_k - \boldsymbol{k}\cdot\boldsymbol{q}/2m. \end{aligned} \tag{3.36}$$

These results are in agreement with those obtained from the Galilei transformation of a system with elementary excitations provided that $\boldsymbol{v}_s$ is set equal to $\boldsymbol{q}/m$ in Eq. (1.18). Furthermore the free energy (3.22) is replaced by

$$F_s = -\sum_k \left\{2\beta^{-1}\ln\left[1+\exp(-\beta\varepsilon_{k+q})\right] + \varepsilon_k - \xi_k + \Delta_k^*\Psi_k\right\}. \tag{3.37}$$

Let us consider the electrical current density first. Since a uniform electrical current is represented by an operator

$$\boldsymbol{J}_s = \frac{e}{m}\sum_{k,\alpha}\boldsymbol{k}a^\dagger_{k\alpha}a_{k\alpha},$$

one needs to evaluate the expectation value of this operator in a superconducting state:

$$\begin{aligned} \boldsymbol{J}_s &= \frac{e}{m}\sum_k\left\{\left(\boldsymbol{k}+\frac{\boldsymbol{q}}{2}\right)\langle a^\dagger_{k_+\uparrow}a_{k_+\uparrow}\rangle + \left(\boldsymbol{k}+\frac{\boldsymbol{q}}{2}\right)\langle a^\dagger_{-k_-\downarrow}a_{-k_-\downarrow}\rangle\right\} \\ &= \frac{e}{m}\sum_k\left\{\boldsymbol{k}\left[f(\varepsilon_{k_+}) - f(\varepsilon_{k_-})\right] + \frac{1}{2}\boldsymbol{q}\left[1 - \frac{\xi_k}{\varepsilon_k} + \frac{\xi_k}{\varepsilon_k}\left(f(\varepsilon_{k_+}) + f(\varepsilon_{k_-})\right)\right]\right\}, \end{aligned} \tag{3.38}$$

where use has been made of the Bogoliubov transformation (3.35). Some of the spin indices are suppressed for simplicity. To obtain the density $n_s(T)$ of the superfluid component, let us keep only terms that are proportional to $\boldsymbol{q}$. Since only the excitations near the Fermi surface contribute to the current, one may put $|\boldsymbol{k}| \simeq k_F$ in the above equation. If $\boldsymbol{q}\sum_k(1-\xi_k/|\xi_k|)/2 = n\boldsymbol{q}/2$ is subtracted from the second term, contributions from the two sides of the Fermi surface $(\pm|\xi|)$ cancel out in this term. Thus one finds

$$\begin{aligned} \boldsymbol{J}_s - ne\boldsymbol{q}/2m &= \frac{1}{m}\sum_k \frac{\partial f(\varepsilon_k)}{\partial\varepsilon_k}\frac{\boldsymbol{k}(\boldsymbol{k}\cdot\boldsymbol{q})}{q} \\ &\simeq \boldsymbol{q}\frac{2}{3}\frac{k_F^2}{m}N(0)\int_0^\infty \mathrm{d}\xi\frac{\partial f(\varepsilon)}{\partial\varepsilon}, \end{aligned} \tag{3.39}$$

where $n = 2N(0)k_F^2/3m$ is the number density. Finally one obtains

$$\boldsymbol{J}_s = n_s(T)e\frac{\boldsymbol{q}}{2m}, \quad n_s(T) = n\left(1 + 2\int_0^\infty \mathrm{d}\xi\frac{\partial f(\varepsilon)}{\partial\varepsilon}\right). \tag{3.40}$$

Note that the second term of $n_s(T)$ is always negative.

Let us consider the case $T = 0$ first, for which $n_s(T) = n$. That is, all the particles form pairs with $\boldsymbol{v}_s = \boldsymbol{q}/m$ and contribute to the supercurrent. For $T \lesssim T_c$, n_s becomes $n(1 - T/T_c)$ and, of course, vanishes at $T = T_c$. The difference $n - n_s(T)$ is of course the normal fluid component made of thermally excited quasi-particles obeying Fermi statistics.

The effect of the centre of mass motion of pairs appears in the gap equation through the distribution function of the excitations. Let us find a term proportional to $\boldsymbol{q}^2|\Delta|^2$, which adds to the free energy F_{sn} near T_c. The function $2f(\varepsilon_k)$ in Eq. (3.29) should be replaced by $f(\varepsilon_k + \boldsymbol{k}\cdot\boldsymbol{q}/2m) + f(\varepsilon_k - \boldsymbol{k}\cdot\boldsymbol{q}/2m)$ for $\boldsymbol{q} \neq 0$. Then it follows that the term proportional to $\boldsymbol{q}^2$ in $(\partial F_{sn}/\partial\Delta^*)\Delta^{-1}$ is

$$\beta^{-1}\sum_k\sum_n\left\{\frac{3}{(\omega_n^2+\varepsilon_k^2)^2} - \frac{4\varepsilon_k^2}{(\omega_n^2+\varepsilon_k^2)^3}\right\}\left(\frac{\boldsymbol{k}\cdot\boldsymbol{q}}{2m}\right)^2.$$

One may put $\varepsilon_k^2 = \xi_k^2$ at $T \simeq T_c$. After carrying out the $\boldsymbol{k}$-integration and the n-summation, one obtains

$$2\beta_c^2 N(0)\frac{7\zeta(3)}{8\pi^2}\frac{1}{3}\frac{k_F^2}{4m^2}\boldsymbol{q}^2|\Delta|^2 = \frac{7\zeta(3)n}{8\pi^2}\beta_c^2\left(\frac{\boldsymbol{q}^2}{4m}\right)|\Delta|^2. \tag{3.41}$$

One has to add the above term to the free energy F_{sn} of Eq. (3.30) when $\boldsymbol{q} \neq 0$. If one substitutes $|\Delta|^2$ of Eq. (3.31) into the above equation, one finds, from Eq. (3.40), that the coefficient of $(\boldsymbol{q}^2/4m)$ is exactly a half of $n_s(T)$, that is, the number of pairs. Problems related to states with superconducting current, such as the critical current in a thin film, will be treated in Chapter 5.

3.2.3 Spin paramagnetism

Let us consider the paramagnetism associated with the electron spin, also known as the Pauli paramagnetism. Suppose the system under consideration is a superconductor under a uniform external magnetic field $\boldsymbol{H}$. Then due to the Meissner effect, the magnetic susceptibility for a uniform $\boldsymbol{B}$ ($\simeq \boldsymbol{H}$) is obtained only for a small particle or a thin film whose thickness is less than the penetration depth, or for a type 2 superconductor near H_{c2}. Note also that only spins couple to a magnetic field in a liquid ^{3}He as we will see in Chapter 6. Let μ_0 be the magnetic moment of an electron. Then the electron energies taking part in the pairing are

$$\xi_{k\uparrow} = \xi_k + \mu_0 H, \quad \xi_{-k\downarrow} = \xi_k - \mu_0 H.$$

This is equivalent to simply replacing $\boldsymbol{k}\cdot\boldsymbol{q}/2m$ by $\mu_0 H$ in Eq. (3.34) obtained for $\boldsymbol{q} \neq 0$. Accordingly the excitation energies are modified as $\varepsilon_{k\uparrow(\downarrow)} =$

$\varepsilon_k + (-)\mu_0 H$. For example the gap equation is now given by

$$1 = N(0)g \int_0^{\omega_c} d\xi \frac{1}{2\varepsilon} \left\{ \tanh \frac{\beta(\varepsilon + \mu_0 H)}{2} + \tanh \frac{\beta(\varepsilon - \mu_0 H)}{2} \right\}. \tag{3.42}$$

Let us consider the case of a weak magnetic field $\mu_0 H \ll \Delta$, which is analogous to the case of small $\boldsymbol{q}$. The magnetisation of the system is obtained as $M = -\partial F_s/\partial H$ from the free energy F_s of Eq. (3.19). In obtaining this, one needs to differentiate the explicit H-dependence of F_s since, although Δ depends on H, the equality $\partial F_s/\partial \Delta = 0$ holds in an equilibrium state. Thus one has

$$M = -\mu_0 \sum_k \{f(\varepsilon_k + \mu_0 H) - f(\varepsilon_k - \mu_0 H)\}.$$

The magnetisation is then given by

$$\chi_s = -2\mu_0^2 \sum_k \frac{\partial f(\varepsilon_k)}{\partial \varepsilon_k} = -2\chi_n \int_0^{\infty} d\xi \frac{\partial f(\varepsilon)}{\partial \varepsilon}.$$

If compared with Eq. (3.40), one finds that it is thermally excited quasiparticles that polarise under a magnetic field. It also follows that

$$\chi_s/\chi_n = \begin{cases} (2\pi\beta\Delta)^{1/2} e^{-\beta\Delta} & (T \ll T_c) \\ 1 - \dfrac{7\zeta(3)}{4\pi^2}\beta^2\Delta^2 & (T \lesssim T_c). \end{cases} \tag{3.43}$$

It can be easily shown from Eq. (3.42) that the magnetic field H, for which Δ vanishes at $T = 0$, satisfies $\mu_0 H/\Delta_0 = 1/2$. At low temperatures, however, the phase transition between the normal state and the superconducting state is of the first order and it takes place at the *Clogston limit* H_P, also known as the *Pauli limit*, defined by

$$\mu_0 H_P/\Delta_0 = 1/\sqrt{2}. \tag{3.44}$$

There are some type 2 superconductors whose H_{c2} obtained in Section 5.3 is larger than H_P. If this is the case, the upper critical field is replaced by H_P.

3.3 Response of a superconducting state to an external field

Suppose the system is interacting with a time-dependent external field $F_i(\boldsymbol{x}, t)$, which introduces an extra term

$$\mathscr{H}_1(t) = \int d\boldsymbol{x}\, C_i(\boldsymbol{x}) F_i(\boldsymbol{x}, t) \tag{3.45}$$

in the Hamiltonian $\mathscr{H}$. F may be an electromagnetic field or an ultrasound wave and can be treated classically in many cases. The subscript i denotes the ith component where F is a vector. This index will be dropped where this does not cause confusion. Suppose the interaction is adiabatically applied to a system which is in equilibrium at $t = -\infty$. This may cause a variation proportional to F in an observable $Q(x)$ of the system, which is known as a linear response. This variation is given by the well-known Kubo formula as

$$\delta Q(\boldsymbol{x},t) = -\mathrm{i}\int\int_{-\infty}^{t} \mathrm{d}\boldsymbol{x}'\mathrm{d}t'\langle[Q(\boldsymbol{x},t), C_i(\boldsymbol{x}',t')]\rangle F_i(\boldsymbol{x}',t'). \tag{3.46}$$

Here operators such as $Q(\boldsymbol{x},t)$ are in the Heisenberg representation whose time evolution is generated by $\mathscr{H} - \mu N$, where μ is the chemical potential and N is the total particle number of the system. The operator Q may be the number density $n(\boldsymbol{x},t)$, the current density $\boldsymbol{j}(\boldsymbol{x},t)$ and so forth. Define the retarded correlation function

$$K_{QC}(\boldsymbol{x}-\boldsymbol{x}',t-t') \equiv -\mathrm{i}\langle[Q(\boldsymbol{x},t), C_i(\boldsymbol{x}',t')]\rangle\theta(t-t') \tag{3.47}$$

for a spatially uniform system. Corresponding to the Fourier decomposition of the external field

$$F(\boldsymbol{x},t) = \sum_{q,\omega} F(\boldsymbol{q},\omega)\exp(\mathrm{i}\boldsymbol{q}\cdot\boldsymbol{x} - \mathrm{i}\omega t),$$

Eqs. (3.46) and (3.47) are decomposed as

$$\begin{aligned}\delta Q(\boldsymbol{q},\omega) &= K_{QC}(\boldsymbol{q},\omega)F(\boldsymbol{q},\omega),\\ K_{QC}(\boldsymbol{q},\omega) &= -\sum_{n,m}\frac{\langle n|Q(\boldsymbol{q})|m\rangle\langle m|C(-\boldsymbol{q})|n\rangle}{\omega - \mathrm{i}\delta - E_m + E_n}\\ &\quad\times Z^{-1}\mathrm{e}^{-\beta(E_n-\mu N)}(1-\mathrm{e}^{-\beta(E_m-E_n)}).\end{aligned} \tag{3.48}$$

Here Z is the partition function, δ is an infinitesimal constant related to the adiabatic switching of the field, and $|n\rangle$ is an energy eigenstate with energy E_n, which is, in the mean field approximation, given by the quasi-particle excitations.

In order to find decay or relaxation of the external field due to the interaction with the system, it is necessary to evaluate the time variation of the total energy of the system

$$\frac{\mathrm{d}\langle(\mathscr{H}+\mathscr{H}_1)\rangle}{\mathrm{d}t} = -\mathrm{i}\int_{-\infty}^{t}\mathrm{d}t'\left\langle\left[\frac{\partial\mathscr{H}_1(t)}{\partial t},\mathscr{H}_1(t')\right]\right\rangle \tag{3.49}$$

and to average it over one period of the external field. Note that only the explicit time dependence of the external field is to be considered in the derivative $\partial\mathscr{H}_1/\partial t$.

The most important cases are those in which $C(\boldsymbol{x})$ and $Q(\boldsymbol{x})$ are the same density such as $n(\boldsymbol{x})$ or $\boldsymbol{j}(\boldsymbol{x})$. Let the Fourier transform of the density be

$$C_j(\boldsymbol{q}) = \sum_{k,\alpha} C_j(\boldsymbol{k}, \boldsymbol{q}) a^\dagger_{k_-\alpha} a_{k_+\alpha}. \tag{3.50}$$

The density C is assumed to be spin independent, although generalisation to a spin-dependent case is straightforward. To evaluate the matrix elements in Eq. (3.48), $C_j(\boldsymbol{q})$ must be expressed in terms of $\gamma^\dagger$ and γ of the Bogoliubov transformation (3.2). (The centre of mass momentum of the pairs, namely $\boldsymbol{q}$ in the previous section, is assumed to be zero. The vector $\boldsymbol{q}$ here denotes the wave vector of the *external* field.) One finds

$$\begin{aligned} C_i(\boldsymbol{q}) = \sum_k C_i(\boldsymbol{k}, \boldsymbol{q}) \{ & (u^*_{k_-} u_{k_+} \mp v^*_{k_+} v_{k_-}) \gamma^\dagger_{k_-\alpha} \gamma_{k_+\alpha} + (v^*_{k_-} v_{k_+} + v^*_{k_+} v_{k_-}) \delta_{q0} \\ & + (u^*_{k_-} v_{k_+} \pm u^*_{k_+} v_{k_-}) \gamma^\dagger_{k_-\uparrow} \gamma^\dagger_{-k_+\downarrow} + (v^*_{k_-} u_{k_+} \pm v^*_{k_+} u_{k_-}) \gamma_{-k_-\downarrow} \gamma_{k_+\uparrow} \}. \end{aligned} \tag{3.51}$$

The signs are chosen according to the parity

$$C_i(-\boldsymbol{k}, \boldsymbol{q}) = \pm C_i(\boldsymbol{k}, \boldsymbol{q}) \tag{3.52}$$

under $\boldsymbol{k} \to -\boldsymbol{k}$. The coefficients of $\gamma^\dagger\gamma, \gamma^\dagger\gamma^\dagger, \gamma\gamma$ in Eq. (3.51) are called *coherence factors* and produce effects characteristic of a superconducting state with a pair formation. Finally one obtains

$$\begin{aligned} K_{C_iC_j}(\boldsymbol{q}, \omega) = & -\sum_k C_i(\boldsymbol{k}, \boldsymbol{q}) C^*_j(\boldsymbol{k}, \boldsymbol{q}) \\ & \times \left[2|(u^*_{k_-} u_{k_+} \mp v^*_{k_+} v_{k_-})|^2 \frac{f(\varepsilon_{k_+}) - f(\varepsilon_{k_-})}{\omega - i\delta + \varepsilon_{k_+} - \varepsilon_{k_-}} \right. \\ & + |(u^*_{k_-} v_{k_+} \pm u^*_{k_+} v_{k_-})|^2 [1 - f(\varepsilon_{k_+}) - f(\varepsilon_{k_-})] \\ & \left. \times \left(\frac{1}{\omega - i\delta - \varepsilon_{k_+} - \varepsilon_{k_-}} - \frac{1}{\omega - i\delta + \varepsilon_{k_+} + \varepsilon_{k_-}} \right) \right] \end{aligned} \tag{3.53}$$

where use has been made of the relation $C^*_i(\boldsymbol{q}) = C_i(-\boldsymbol{q})$, which follows since $C(\boldsymbol{x})$ is real. The first term represents processes in which thermally excited quasi-particles are scattered by the external field while the second term corresponds to quasi-particle pair creations/annihilations due to the external field. The coherence factors are obtained, using Eq. (3.3), as

$$|u_{k_-}^* u_{k_+} \mp v_{k_+}^* v_{k_-}|^2 = \frac{1}{2\varepsilon_{k_-}\varepsilon_{k_+}}(\varepsilon_{k_-}\varepsilon_{k_+} + \xi_{k_-}\xi_{k_+} \mp |\Delta|^2)$$

$$|u_{k_-}^* v_{k_+} \pm u_{k_+}^* v_{k_-}|^2 = \frac{1}{2\varepsilon_{k_-}\varepsilon_{k_+}}(\varepsilon_{k_-}\varepsilon_{k_+} - \xi_{k_-}\xi_{k_+} \pm |\Delta|^2). \tag{3.54}$$

Let us now consider some applications.

3.3.1 Ultrasound absorption

The simplest application is to ultrasound absorption. This is because the wave number $\boldsymbol{q}$ is uniquely specified by the given frequency ω, once the wave mode is chosen. This is in contrast with the electromagnetic field discussed later. Furthermore the sound velocity $v_s = \omega/q$ is much smaller than the Fermi velocity v_F. For the longitudinal wave considered in the following, the quantity $C(\boldsymbol{x})$ in Eq. (3.45) is the particle density $n(\boldsymbol{x})$ and one has $C(\boldsymbol{k}, \boldsymbol{q}) = 1$. Let $\phi_\omega e^{i(\boldsymbol{q}\cdot\boldsymbol{x}-\omega t)}$ be the potential for electrons due to the sound wave (a lattice wave of a long wavelength). It can be shown from Eq. (3.49) that the sound wave absorption is given by

$$\omega\ \mathrm{Im}K(\boldsymbol{q}, \omega)|\phi_\omega|^2/2.$$

Here $K(\boldsymbol{q}, \omega)$ is given by Eq. (3.53) with the substitution $C_i = 1$. Since an ultrasound frequency is usually much smaller than $|\Delta(T)|$, only the first term of Eq. (3.53) contributes to the imaginary part and one obtains

$$\mathrm{Im}\,K(\boldsymbol{q}, \omega) = -\pi \sum_k \left(1 + \frac{\xi_+\xi_- - |\Delta|^2}{\varepsilon_+\varepsilon_-}\right) \times [f(\varepsilon_+) - f(\varepsilon_-)]\delta(\omega - \varepsilon_- + \varepsilon_+).$$

Note the appearance of the coherence factor. Let us make the change of variables

$$\sum_k \cdots \to N(0)\frac{1}{4qv_F}\int\int d\xi_+ d\xi_- \tag{3.55}$$

which is obtained by using $\xi_+ - \xi_- = \hat{\boldsymbol{k}}\cdot\boldsymbol{q}v_F$ (note that $qv_F \gg \omega = qv_s$) and $\xi = (\xi_+ + \xi_-)/2$. Since $\omega \ll |\Delta|$ one finally obtains

$$\mathrm{Im}K(\boldsymbol{q}, \omega) \propto \omega \int_\Delta^\infty d\varepsilon \left(\frac{\varepsilon}{\xi}\right)^2 \left(1 - \frac{|\Delta|^2}{\varepsilon^2}\right) \frac{\partial f}{\partial \varepsilon}. \tag{3.56}$$

Note that the density of states and the coherence factor exactly cancel out and the integration simply gives a function $f(\Delta)$. In comparing theory

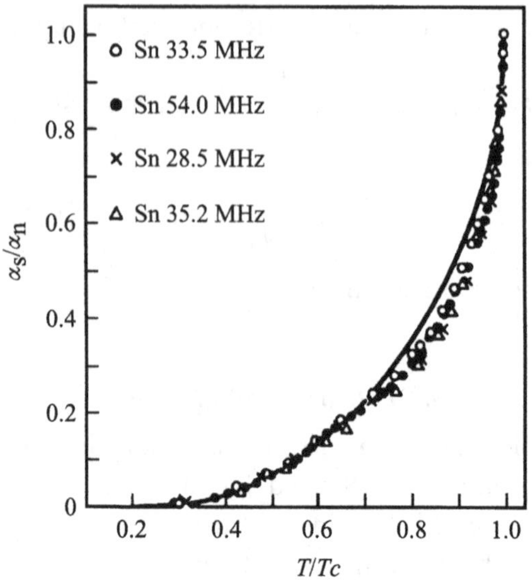

Fig. 3.4. Ultrasound absorption. (From [G-6].)

with experiment, it turns out to be convenient to normalise the absorption coefficient relative to the normal state value as

$$\frac{\alpha_s}{\alpha_n} = 2f\left(\Delta(T)\right). \tag{3.57}$$

Since $v_F \gg v_s$, sound absorption takes place when quasi-particles near the equator of the Fermi surface with momenta perpendicular to $\boldsymbol{q}$ are scattered by the sound wave. Therefore $\Delta(T)$ in Eq. (3.57) is the energy gap averaged over the equator. As is shown in Fig. 3.4, α_s/α_n directly measures the temperature dependence of the gap. It should be added that in the case of transverse waves the absorption suddenly becomes small as the temperature is reduced below T_c. This is due to the Meissner effect since, in the free electron model, the vector potential appears in the absorption coefficient. In reality, however, one must also take into account the finite electron mean free path l and electron band effect.

3.3.2 Response to electromagnetic field

Let $\boldsymbol{A}(\boldsymbol{x}, t)$ be the vector potential of an electromagnetic field. Here the gauge $\nabla \cdot \boldsymbol{A} = 0$ is chosen for reasons which will become clear later. According to quantum mechanics, the electronic current density operator is

$$j(x) = \frac{e}{2m}\sum_{\alpha}\left\{\psi_{\alpha}^{\dagger}(x)\left(\frac{\nabla}{\mathrm{i}} - \frac{e}{c}A\right)\psi_{\alpha}(x) + \left(-\frac{\nabla}{\mathrm{i}} - \frac{e}{c}A\right)\psi_{\alpha}(x)\cdot\psi_{\alpha}(x)\right\}$$

$$= j_{\mathrm{p}}(x) - \frac{\mathrm{e}^2}{mc}n(x)A(x,t). \tag{3.58}$$

Here the first term is called the paramagnetic current and the second term the diamagnetic current. The interaction Hamiltonian which is proportional to A is

$$\mathscr{H}_1 = -\frac{1}{c}\int \mathrm{d}x\, j_{\mathrm{p}}(x)A(x,t). \tag{3.59}$$

Let us compute the electronic current density itself as a response to the applied field.

Since the second term in Eq. (3.58) is already proportional to A, one may replace $n(x)$ by the mean particle density n. The first term

$$j_{\mathrm{p}}(q) = \sum_{k\ \alpha}\frac{e}{m}k a_{k_-\alpha}^{\dagger}a_{k_+\alpha}$$

is obtained by taking $C_i(k,q) = ek_i/m$ in Eq. (3.53).

Since general cases are complicated, let us concentrate on the case $T = 0$ here. The response in this case is given by

$$\langle j_i(q,\omega)\rangle = \left[-\frac{ne^2}{mc} + K(q,\omega)\right]A_i(q,\omega)$$

$$K(q,\omega) = -\frac{\mathrm{e}^2}{2m^2c}\sum_k\left(k^2 - \frac{(k\cdot q)^2}{q^2}\right)\left(1 - \frac{\xi_{k_-}\xi_{k_+} + |\Delta|^2}{\varepsilon_{k_-}\varepsilon_{k_+}}\right)$$

$$\times\left(\frac{1}{\omega - \mathrm{i}\delta - \varepsilon_{k_-} - \varepsilon_{k_+}} - \frac{1}{\omega - \mathrm{i}\delta + \varepsilon_{k_-} + \varepsilon_{k_+}}\right). \tag{3.60}$$

Let us consider a static magnetic field, $\omega = 0$, first. It can be easily shown that $K(q,\omega) \simeq O((qv_{\mathrm{F}}/\Delta_0)^2)$ provided that $qv_{\mathrm{F}} \ll |\Delta_0|$, namely $q\xi_0 \ll 1$ (the *London limit*). In this limit, one obtains the *London equation*,

$$j = -\frac{ne^2}{mc}A.$$

As was shown in Chapter 1, the Meissner effect can be derived if the above equation is solved simultaneously with the Maxwell equations. Note that if one takes $\Delta = 0$ in Eq. (3.60), one obtains $K = ne^2/mc$, which exactly cancels the first term of $\langle j_i(q,\omega)\rangle$. In a normal state, no current proportional to the vector potential A exists and, as in the Ohmic law, the lowest order contribution to the current response is from terms proportional to ωA.

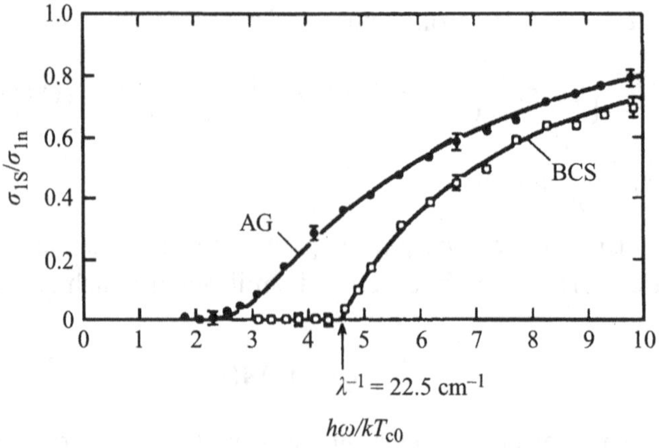

Fig. 3.5. The ω-dependence of σ_1 for pure lead (□) and lead with a magnetic impurity Gd (•). See Section 4.6 for the AG theory. (From [G-7].)

Let us next consider the imaginary part of Eq. (3.60). In contrast to ultrasound absorption, electromagnetic wave absorption is quite complicated, since $A(\boldsymbol{q}, \omega)$ in a sample is obtained by simultaneously solving the Maxwell equations with Eq. (3.60). Accordingly when the incident wave has a finite frequency ω, one generally has skin effect. In a superconductor there is also the Meissner effect. Hence the wave penetrates into the sample within a finite depth δ_{P}. Therefore one has to calculate $K(\boldsymbol{q}, \omega)$ for $|\boldsymbol{q}| \sim \delta_{\mathrm{P}}^{-1}$. For a thin film with thickness d less than the coherence length ($d \ll \xi = v_{\mathrm{F}}/\Delta$), however, the $\boldsymbol{q}$-dependence is negligible, which makes the analysis easier. The imaginary part is obtained by putting $\mathrm{Im}(\omega - \mathrm{i}\delta - \varepsilon_{k_-} - \varepsilon_{k_+})^{-1} = \pi\delta(\omega - \varepsilon_{k_-} - \varepsilon_{k_+})$ in Eq. (3.60). One may make the replacement (3.55) here as well since $qv_{\mathrm{F}} \gg \Delta_0 \sim \omega$. The result is the expression for the real part (also known as the resistive part) $\sigma_1(\omega)$ of the ω-dependent electrical conductivity $\sigma_1(\omega) + \mathrm{i}\sigma_2(\omega) = -\mathrm{i}(c/\omega)K(\omega)$. As in ultrasound attenuation, we are interested in the ratio normalised by the normal conductivity,

$$\frac{\sigma_{1\mathrm{s}}}{\sigma_{1\mathrm{n}}} = \frac{1}{\omega}\int_{\Delta}^{\omega-\Delta} \mathrm{d}\varepsilon \frac{\varepsilon(\varepsilon-\omega) - \Delta_0^2}{(\varepsilon^2 - \Delta_0^2)^{1/2}((\omega-\varepsilon)^2 - \Delta_0^2)^{1/2}}. \tag{3.61}$$

The value of $\sigma_{1\mathrm{s}}/\sigma_{1\mathrm{n}}$ measured with a thin film of lead is in good agreement with this prediction based on the BCS theory. Observe, in particular, that absorption starts when ω exceeds the energy gap, see Fig. 3.5.

Analogous to the electrical conductivity, the thermal conductivity κ is also an important transport coefficient. However, here we are content with the following comments. The contribution to the thermal conductivity in a normal metal is mainly from the conduction electrons and there exists the Wiedemann–Franz relation $\kappa = (\pi^2/3)(k_B/e)^2 T\sigma$ between the electrical conductivity $\sigma = 1/\rho$ and the thermal conductivity κ. In a superconducting state, however, σ is infinite while κ becomes less than that in a normal state. This is because the current is carried by the pair condensate while the entropy is carried by the gas of thermally excited quasi-particles only. At a low temperature $T \ll T_c$, the thermal conductivity κ is small, proportional to $\exp(-\beta\Delta_0)$ (see [C-2] vol. I, [C-3]).

Gauge invariance and collective modes

The above formulation, in which the electromagnetic response has been calculated by taking quasi-particle excitations only into account, is valid only in the gauge $\nabla \cdot \boldsymbol{A} = 0$. In other words, the formulation is not invariant under the gauge transformation $\boldsymbol{A} \to \boldsymbol{A} + \nabla\chi(\boldsymbol{x})$. In quantum mechanics, the state vector is multiplied by a phase factor $\exp(\mathrm{i}N\chi)$, N being the total number of particles, at the same time as gauge transformation of the vector potential. Accordingly the order parameter $\Psi = \langle N-2|\psi\psi|N\rangle$ must change by a phase $\exp(\mathrm{i}2\chi)$. Then the phase of Δ and hence that of v_k changes by the same amount; this has been neglected in the above formalism, and, accordingly, the result is not gauge-invariant. It has been noted in Chapter 1 that the symmetry broken in a superconducting or superfluid state is the invariance under gauge transformation due to the appearance of condensate with a definite phase. It is impossible, however, that the final result depends on an arbitrarily chosen gauge. To be specific, there cannot be a finite current with the choice of a gauge $\boldsymbol{A} = \nabla\chi$. This difficulty disappears if a collective mode called the Bogoliubov–Anderson mode, that is, oscillation of the phase of the order parameter, is considered. The Bogoliubov–Anderson mode is an example of the so-called Nambu–Goldstone mode associated with spontaneous breaking of symmetries.

If $|\Psi_0\rangle$ denotes the BCS ground state with a condensate of the form $(\boldsymbol{k}\uparrow, -\boldsymbol{k}\downarrow)$, the Bogoliubov–Anderson mode is an excited state given approximately by

$$\sum_k f(\boldsymbol{k},\boldsymbol{q})(a^\dagger_{k_+\uparrow}a^\dagger_{-k_-\downarrow} - a_{-k_-\downarrow}a_{k_+\uparrow})|\Psi_0\rangle. \tag{3.62}$$

In the limit $\boldsymbol{q} \to 0$, this is simply equivalent to shifting the phase of the pair by a constant amount and the energy remains the same as that of the ground state. It can be shown that the electrical current response against the longitudinal component of the vector potential vanishes if this excitation is taken into account. (One has to use the Ward–Takahashi relation to construct a rigorous theory with gauge invariance, see [C-3].)

If the phase χ of the pair varies spatially as $\boldsymbol{q} \cdot \boldsymbol{x}$, there exists a uniform superflow as was shown in Section 3.2. If the phase χ changes as $e^{i\boldsymbol{q}\cdot\boldsymbol{x}}$, then the superflow is not uniform and hence there exist density fluctuations. Consequently this collective mode with finite q couples with density fluctuations. In an electron system this excitation becomes a plasma oscillation with a large frequency due to the Coulomb interaction. Accordingly this excitation does not directly influence such phenomena as superconductivity.

Collective modes play very interesting rôles in a superconductor or a superfluid with a pair condensate with nonvanishing angular momentum ($l \neq 0$) that is, other than s-wave pairing, see Chapter 6.

3.3.3 Nuclear magnetic resonance

The nucleus of many metals that show superconductivity has a spin. Let $\boldsymbol{I}$ denote the nuclear spin and $\boldsymbol{m}_N = \gamma_N \boldsymbol{I}$ the associated magnetic moment. This moment precesses in a static magnetic field $\boldsymbol{H}$. The resonance frequency shifts from $\omega_0 = \gamma_N H$ and, moreover, the precession decays due to interaction with the surroundings. In a metal, the moment interacts with the conduction electrons and information on the latter is obtained through nuclear magnetic resonance. The origin of the interaction is the coupling of the nuclear magnetic moment with the electron magnetic moments associated with the electron spin and the orbital angular momentum. We consider only the electron spins here to simplify the discussion. The magnetic field due to a magnetic moment $\boldsymbol{m}_N$ placed at the origin is given by

$$\boldsymbol{B}(\boldsymbol{x}) = \frac{3\boldsymbol{x}(\boldsymbol{x} \cdot \boldsymbol{m}_N) - |\boldsymbol{x}|^2 \boldsymbol{m}_N}{|\boldsymbol{x}|^5} + \frac{8\pi}{3} \boldsymbol{m}_N \delta(\boldsymbol{x}).$$

Hence the interaction between $\boldsymbol{m}_N$ and a magnetic moment associated with the electron spin $\boldsymbol{s}$ is

$$\begin{aligned} \mathscr{H}_{sI} &= -\mu_0 \int \mathrm{d}\boldsymbol{x}\, \boldsymbol{s}(\boldsymbol{x}) \cdot \boldsymbol{B}(\boldsymbol{x}) \\ \boldsymbol{s}(\boldsymbol{x}) &= \psi^*_\alpha(\boldsymbol{x}) \boldsymbol{s}_{\alpha\beta} \psi_\beta(\boldsymbol{x}). \end{aligned} \tag{3.63}$$

Suppose the electrons belong to the s-band and the probability $|\psi(\boldsymbol{x})|^2$ is isotropic around the nucleus. Then only the δ-function term contributes to the interaction. Therefore if $\boldsymbol{R}_i$ denotes the lattice points on which the nuclei sit, the interaction between an electron and a nucleus is given by

$$\mathcal{H}_1 = -\frac{8\pi}{3}\mu_0 \int \mathrm{d}\boldsymbol{x}\, \boldsymbol{s}(\boldsymbol{x}) \cdot \sum_i \gamma_\mathrm{N} \boldsymbol{I}_i \delta(\boldsymbol{x} - \boldsymbol{R}_i). \tag{3.64}$$

Shift of the resonance frequency (Knight shift)

Since ω_0 is small compared to the electron spin frequency, the nuclear magnetic moment experiences an averaged electron spin. The spin part of the magnetisation of the electron system in an external magnetic field $\boldsymbol{H}$ is

$$M = \mu_0 \left\langle \int \mathrm{d}\boldsymbol{x}\, \boldsymbol{s}(\boldsymbol{x}) \right\rangle = \chi H,$$

where χ is the magnetic susceptibility. Accordingly each spin is under an extra magnetic field (called the hyperfine field) of $(8\pi/3)|\psi(0)|^2\chi \boldsymbol{H}$ according to Eq. (3.64). Here $\psi(0)$ is the electron wave function at the position of the nuclei. Therefore the shift of the resonance frequency is given by

$$K \equiv \frac{\omega - \omega_0}{\omega_0} = \frac{8\pi}{3}|\psi(0)|^2\chi \tag{3.65}$$

and is called the *Knight shift*. Although it is rather difficult to fix $|\psi(0)|^2$ quantitatively, the ratio

$$K_\mathrm{s}/K_\mathrm{n} = \chi_\mathrm{s}/\chi_\mathrm{n}$$

may be compared with the theory. According to Eq. (3.43) the ratio $\chi_\mathrm{s}/\chi_\mathrm{n} \to 0$ as $T \to 0$. Figure 3.6 shows the Knight shifts for Al, Hg and Sn. The shift for Al is close to the theoretical curve while those for Sn and Hg do not vanish as $T \to 0$. This discrepancy is attributed to spin–orbit scattering ($S - L$ scattering). The samples consist of fine particles since microwaves are used in actual experiments, for which $S - L$ scattering at the surface is not negligible. The electron spin is no longer a good quantum number in the presence of $S - L$ scattering. Accordingly the pair cannot remain in the pure spin-singlet state. As a result, the Knight shift χ_s generally remains finite as $T \to 0$. The Knight shift tends to vanish as $T \to 0$ for such metals as Al, for which the contribution from $S - L$ scattering is small. There is an additional contribution χ_orb to the magnetisation due to the orbital motion of electrons. This may be separated from other contributions since it has minimal temperature dependence.

Nuclear spin relaxation time T_1

Let us calculate the transition probability with which each nuclear spin undergoes a transition from a state with energy ω_0 to that with vanishing

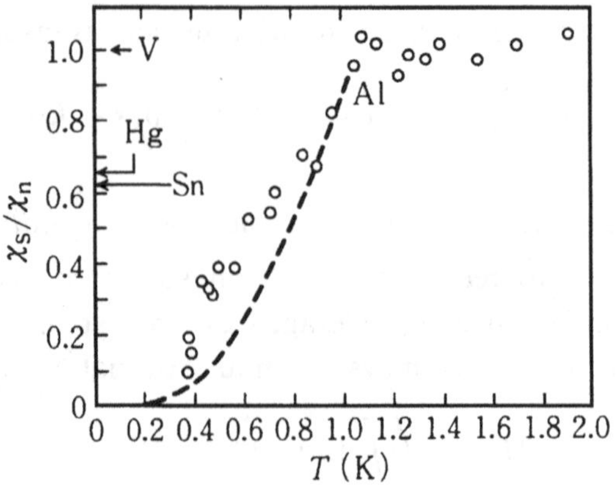

Fig. 3.6. The Knight shift of Al. The arrows in the figure depict the shifts for Hg and Sn in the limit $T \to 0$. They do not fall to zero because of $S - L$ scattering. (From [G-8].)

energy, induced by spin–electron interaction (3.64), assuming that there are no correlations among motions of different nuclear spins. If one concentrates on a particular nuclear spin, the interaction between the spin and conduction electrons is given by $\mathscr{H}_1 \propto \sum_{k,q}\left(a^\dagger_{k_+\uparrow}a_{k_-\downarrow}I^{(-)} + \text{h.c.}\right)$, where $I^{(\pm)} \equiv I_x \pm \mathrm{i}I_y$. The second order perturbation in $\mathscr{H}_1$ gives the transition probability

$$T_{1\mathrm{s}}^{-1} \propto \sum_{k,q} f(\varepsilon_{k_+})[1 - f(\varepsilon_{k_-})]|u^*_{k_-}u_{k_+} + v^*_{k_+}v_{k_-}|^2\delta(\varepsilon_{k_+} + \omega - \varepsilon_{k_-}).$$

Quasi-particle pair creations are neglected in the above equation since the inequality $\omega \ll \Delta$ holds as before. Note that the coherence factor in Eq. (3.51) is odd in this case and that there is $\boldsymbol{q}$-summation since the scattering processes occur at a nuclear site. For an isotropic model, one has

$$\begin{aligned} T_{1\mathrm{s}}^{-1} &= \int\!\!\int \mathrm{d}\xi_+\mathrm{d}\xi_- f(\varepsilon_+)[1 - f(\varepsilon_-)]\left(1 + \frac{\xi_+\xi_- + \Delta^2}{\varepsilon_+\varepsilon_-}\right)\delta(\varepsilon_+ + \omega - \varepsilon_-) \\ &= 4\int_\Delta^\infty \mathrm{d}\varepsilon \frac{\varepsilon}{\sqrt{\varepsilon^2 - \Delta^2}}\frac{\varepsilon + \omega}{\sqrt{(\varepsilon + \omega)^2 - \Delta^2}} \\ &\qquad \times f(\varepsilon + \omega)[1 - f(\varepsilon)]\left(1 + \frac{\Delta^2}{\varepsilon(\varepsilon + \omega)}\right). \end{aligned} \tag{3.66}$$

The contributions above and below the Fermi surface cancel in the term proportional to $\xi_+\xi_-$. If one puts $\Delta = 0$ in the above equation, it reduces to

$$\int_0^\infty \mathrm{d}\xi f(\xi)[1 - f(\xi)] = \frac{1}{2}k_\mathrm{B}T.$$

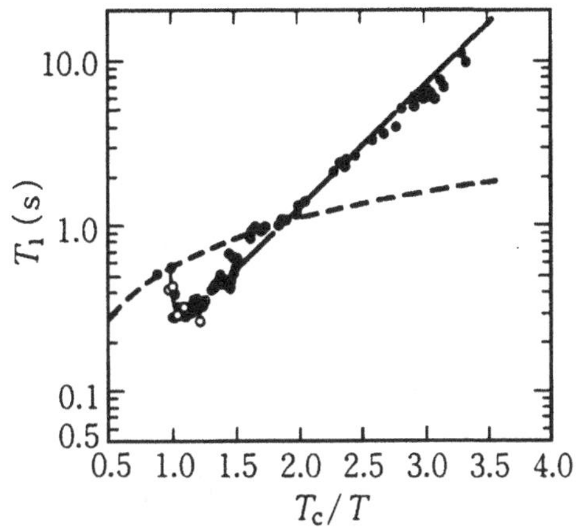

Fig. 3.7. The temperature dependence of T_1 for ^{27}Al. The solid line is the theoretical curve in which the anisotropy of the gap is taken into account. The dotted line denotes T_1 for a normal state. (The dots • are taken from [G-9], while ∘ are from [G-10].)

If the proportionality constant is supplied, one obtains the *Korringa equation*

$$T_{1n}^{-1} = \left(\frac{8\pi}{3}\mu_0\gamma_N|\psi(0)|^2 N(0)\right)^2 \pi k_B T \tag{3.67}$$

in the normal state. It should be noted that T_1^{-1} is proportional to T as a result of the Fermi degeneracy. The coherence factor and the density of states, as well as the Fermi distribution functions, appear in Eq. (3.66). As a result, the integrand diverges logarithmically at the lower limit $\varepsilon = \Delta$ for $\omega \sim 0$. Accordingly it is expected that T_1^{-1} becomes very large below T_c. In actual metals, however, the electron–phonon interaction, which is the origin of superconductivity, and also the density of states are not isotropic. As a result, the energy gap has an anisotropy of up to 10% on the Fermi surface. The logarithmic divergence is weakened once the gap is averaged over the Fermi surface. Taking this into account, the theoretical prediction agrees with experiment, see Fig. 3.7. It should be added that the strong coupling effect treated in Chapter 4 also tends to weaken the increase of T_{1s}^{-1}. In any case, the increase of T_1^{-1} just below T_c should be contrasted with the simple suppression of the ultrasound absorption coefficient α_s. Equation (3.66) becomes exponentially small at low temperatures ($\Delta(T) \sim \Delta_0$) due to the factors containing the distribution function, which can be seen in Fig. 3.7. (See [E-1] for a review of nuclear magnetic resonance.)

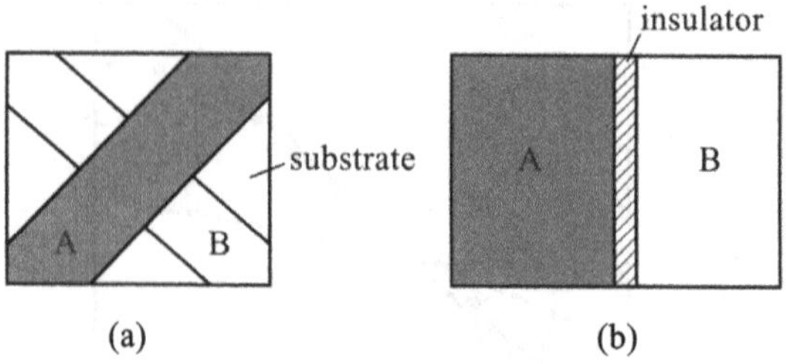

Fig. 3.8. Tunnelling junctions.

3.4 Tunnelling junction and Josephson effect

The tunnelling junction is one of the most powerful phenomena for the study of superconducting states. Firstly, it makes possible direct observation of the density of states associated with the energy gap. Similarly, the Josephson effect due to pair tunnelling reveals the rôle played by the phase of the order parameter, that is, symmetry breaking associated with superconductivity. Secondly, the tunnelling characteristic shows in detail how the electron–phonon interaction leads to pair formation in ordinary superconductors. The latter will be studied in Chapter 4. Let us consider here the tunnelling effect in a superconductor within the BCS model.

A typical tunnelling junction is made of two metals in contact through a thin insulator (an oxide film in many cases, one can even use a normal metal) as shown in Fig. 3.8. The contact may be a weak one such as a point contact made by a needle. Figure 3.8(b) is a model of a junction in which metals A and B couple through a tunnelling barrier. Let us denote the state of an electron in the side A (B) by $\boldsymbol{k}, \alpha$ $(\boldsymbol{p}, \beta)$. An electron transfer from one side to the other is possible since there is an overlap of electron wave functions inside the barrier. Let us introduce the tunnelling Hamiltonian

$$H_{\mathrm{T}} = \sum_{k,p,\alpha} T_{kp} a^{\dagger}_{k\alpha} c_{p\alpha} + \text{h.c.}, \tag{3.68}$$

which couples A and B, to describe the effect. Here $a_{k\alpha}(c_{p\alpha})$ is an electron operator of the system A (B) and T_{kp} is a matrix element which transfers an electron $\boldsymbol{p}, \alpha$ in B to an electron $\boldsymbol{k}, \alpha$ in A, which is assumed to be independent of the spin. The matrix element satisfies $T_{kp} = T^{*}_{-k,-p}$ due to the time-reversal invariance. This is the requirement that the probability amplitude for the time-reversed transfer be the same as before the time-reversal. Processes accompanying phonons or those mediated by magnetic impurities are of course neglected in H_{T} of Eq. (3.68). The tunnelling current is the derivative of the number of electrons in the system A with respect to time, which is given by the commutation relation of $N_{\mathrm{A}} = \sum_{k,\alpha} a^{\dagger}_{k\alpha} a_{k\alpha}$ and H_{T}, namely

$$J = \mathrm{i}e \sum_{k,p,\alpha} [T_{kp} a^{\dagger}_{k\alpha} c_{p\alpha} - \mathrm{h.c.}]. \tag{3.69}$$

In most cases, a constant voltage V is applied between the two sides of a tunnelling junction. If system A has an electric potential $-eV$ while B has 0 potential, electrons in A have more energy than those in B by $-eV$. Accordingly the matrix element $\langle m | a^{\dagger}_{k\alpha} c_{p\alpha} | n \rangle$, which transfers an electron from A to B, has an extra phase $\mathrm{e}^{-\mathrm{i}eVt}$ in addition to the ordinary phase $\mathrm{e}^{-\mathrm{i}(E_n - E_m)t}$. Here $|m\rangle$ and $|n\rangle$ are energy eigenstates of the system A + B in the absence of H_{T}. It is convenient to include this extra phase in H_{T} from the beginning. Let us define

$$H_{\mathrm{T}}(t) = h_{\mathrm{T}} \mathrm{e}^{-\mathrm{i}eVt} + h_{\mathrm{T}}{}^{\dagger} \mathrm{e}^{\mathrm{i}eVt}, \quad h_{\mathrm{T}} \equiv \sum_{k,p,\alpha} T_{kp} a^{\dagger}_{k\alpha} c_{p\alpha} \tag{3.70}$$

for this purpose. We need to evaluate the expectation value of the current J in the presence of H_{T}. The tunnelling junction is assumed to be a weak interaction and hence we may keep lowest order terms, that is, terms of the order of $|T|^2$. This is just the linear response theory introduced in Section 3.3 and we simply adopt the formalism developed there. From Eqs. (3.69) and (3.70) one finds

$$\begin{aligned} \langle J(t) \rangle &= -\mathrm{i} \int_{-\infty}^{t} \mathrm{d}t' \langle [J(t), H_{\mathrm{T}}(t')] \rangle \\ &= 2e \,\mathrm{Re} \int_{-\infty}^{t} \mathrm{d}t' \{ \langle [h_{\mathrm{T}}(t), h_{\mathrm{T}}{}^{\dagger}(t')] \rangle \mathrm{e}^{-\mathrm{i}eV(t-t')} \\ &\quad + \langle [h_{\mathrm{T}}(t), h_{\mathrm{T}}(t')] \rangle \mathrm{e}^{-\mathrm{i}eV(t+t')} \}. \end{aligned} \tag{3.71}$$

The second term is characteristic of a superconductor and it yields a finite contribution because the electron number is not a good quantum number in each system, A and B. If Eq. (3.71) is written in the same form as Eq. (3.48), one obtains

$$\begin{aligned} \langle J(t) \rangle = &-2e \,\mathrm{Im} \sum_{n,m} Z^{-1} (\mathrm{e}^{-\beta E_n} - \mathrm{e}^{-\beta E_m}) \frac{1}{E_m - E_n + eV - \mathrm{i}\delta} \\ &\times \{ |\langle n | h_{\mathrm{T}} | m \rangle|^2 + \mathrm{e}^{-\mathrm{i}2eVt} \langle n | h_{\mathrm{T}} | m \rangle \langle m | h_{\mathrm{T}} | n \rangle \}. \end{aligned} \tag{3.72}$$

3.4.1 Quasi-particle term

The summation over states in Eq. (3.72) is carried out separately for system A and system B. Let us introduce the *spectral density* of each system as

$$\rho_{A\alpha}{}^{(G)}(\boldsymbol{k},\omega) \equiv \sum_{n_A,m_A} Z_A{}^{-1} \exp(-\beta E_{n_A}) |\langle n_A | a_{k\alpha} | m_A \rangle|^2 \\ \times \delta(\omega - E_{m_A} + E_{n_A})(1 + e^{-\beta\omega}) \tag{3.73}$$

and similarly for $\rho_{B\alpha}{}^{(G)}(\boldsymbol{p},\omega)$. The contribution from the first term in the curly brackets in Eq. (3.72) can be written in terms of these spectral densities as

$$I = -2e \sum_{k,p} |T_{kp}|^2 \iint d\omega_1 d\omega_2 \, \pi\delta(\omega_2 - \omega_1 + eV) \\ \times \rho_{A\alpha}{}^{(G)}(\boldsymbol{k},\omega_1)\rho_{B\alpha}{}^{(G)}(\boldsymbol{p},\omega_2)[f(\omega_1) - f(\omega_2)]. \tag{3.74}$$

Suppose $|T_{kp}|^2$ may be replaced by the averaged value over the states near the Fermi surface, which is a good approximation in most cases. Then the above equation becomes

$$I = -4\pi e \langle |T_{kp}|^2 \rangle \int d\omega \mathscr{D}_A(\omega) \mathscr{D}_B(\omega - eV)[f(\omega) - f(\omega - eV)], \tag{3.75}$$

where

$$\mathscr{D}_A(\omega) = \sum_{k,\alpha} \rho_{A\alpha}{}^{(G)}(\boldsymbol{k},\omega) \tag{3.76}$$

is the density of states.

Since states in the BCS model are described by quasi-particle excitations, the spectral density Eq. (3.73) is readily obtained as

$$\rho^{(G)}(\boldsymbol{k},\omega) = |u_k|^2 \delta(\omega - \varepsilon_k) + |v_k|^2 \delta(\omega + \varepsilon_k). \tag{3.77}$$

If, furthermore, there exists the particle–hole symmetry, the coherence factor vanishes upon the summation in Eq. (3.76) and the density of states becomes

$$\mathscr{D}(\omega) = N(0) \frac{|\omega|}{\sqrt{\omega^2 - |\Delta|^2}} \theta(|\omega| - |\Delta|),$$

which is the density of states given by Eq. (3.7).

1. I_{nn}. If both A and B are in the normal state, the densities of states are $\mathscr{D}_A(\omega) = N_A(0)$ and similarly for B. Then one readily finds

$$I_{nn} = R^{-1} V \\ R^{-1} = 4\pi e^2 \langle |T_{kp}|^2 \rangle N_A(0) N_B(0). \tag{3.78}$$

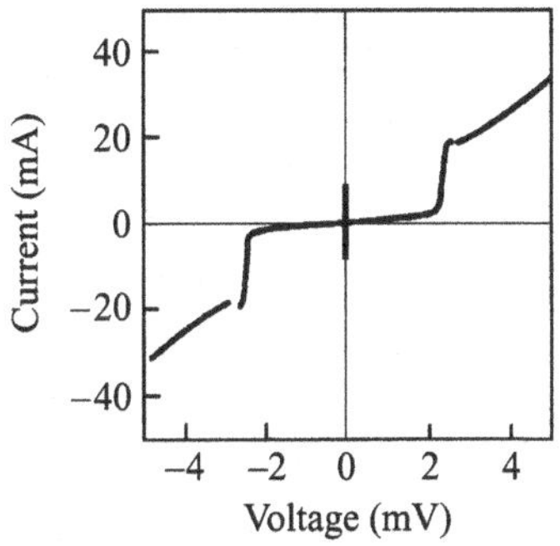

Fig. 3.9. The $I - V$ characteristic curve of a Nb–Pb tunnelling junction at $T = 1.4$ K. Note the Josephson current at $V = 0$.

The tunnelling current obeys Ohm's law and R^{-1} is the conductance per unit area.

2. I_{sn}. Suppose the system B is in the normal state. This case is important in studying the electron–phonon interaction for example. To compare theory with experiment, it is convenient to analyse $\mathrm{d}I_{\mathrm{sn}}/\mathrm{d}V$:

$$\frac{\mathrm{d}I_{\mathrm{sn}}}{\mathrm{d}V} = (N_{\mathrm{A}}(0)R)^{-1} \int_{-\infty}^{\infty} \mathrm{d}\omega \mathscr{D}_{\mathrm{A}}(\omega) \frac{\partial f(\omega - eV)}{\partial \omega}. \tag{3.79}$$

One may measure the density of states $\mathscr{D}_{\mathrm{A}}(\omega)$ since the derivative of the distribution function is finite with width $k_{\mathrm{B}}T$ near $\omega = eV$. An explicit example will be given in Chapter 4.

3. Figure 3.9 shows the tunnelling characteristic I_{ss} for which both sides of the junction are superconductors. This case is studied in detail in the next subsection.

3.4.2 Josephson term

Let J be the contribution from the second term in the curly brackets in Eq. (3.72). In order to write this term in a form similar to Eq. (3.74), we introduce a new spectral density

$$\rho^{(F)}(\boldsymbol{k}, \omega) = -\sum_{n,m} Z^{-1} \mathrm{e}^{-\beta E_n} \langle n | a_{k\uparrow} | m \rangle \langle m | a_{-k\downarrow} | n \rangle \delta(\omega - E_m + E_n)(1 + \mathrm{e}^{-\beta\omega}), \tag{3.80}$$

where the subscript A or B has been dropped to simplify the notation. It readily follows from Eq. (3.2) that the new spectral density in the BCS model is given by

$$\rho^{(F)}(\boldsymbol{k}, \omega) = +u_k v_k [\delta(\omega - \varepsilon_k) - \delta(\omega + \varepsilon_k)]. \tag{3.81}$$

If $|T_{kp}|^2$ is replaced by the averaged value as before, one has the expression corresponding to Eq. (3.74),

$$J = -\frac{1}{2}(eRN_A(0)N_B(0))^{-1}\iint d\omega_1 d\omega_2 \,\mathrm{Im}\left(\frac{1}{\omega_2-\omega_1+eV-i\delta}e^{-i2eVt}\right.$$
$$\left.\times\sum_{k,p}\rho_A{}^{(F)*}(\boldsymbol{k},\omega_1)\rho_B{}^{(F)*}(\boldsymbol{p},\omega_2)\right)[f(\omega_1)-f(\omega_2)]. \quad (3.82)$$

Let us recall here that the factor $u_k v_k$ in Eq. (3.81) is proportional to the order parameter. Accordingly $\sum_k \rho_A^{(F)}(\boldsymbol{k},\omega)$ is proportional to the order parameter, that is, the wave function of the condensate, of the system A. In the simple BCS model one therefore has

$$\sum_k \rho^{(F)}(\boldsymbol{k},\omega) \equiv e^{i\chi}\tilde{\mathscr{D}}(\omega) = e^{i\chi}\sum_k \frac{|\Delta|}{2\varepsilon_k}[\delta(\omega-\varepsilon_k)-\delta(\omega+\varepsilon_k)]$$
$$= e^{i\chi}N(0)\frac{\omega}{|\omega|}\frac{|\Delta|}{\sqrt{\omega^2-|\Delta|^2}}\theta(\omega^2-|\Delta|^2). \quad (3.83)$$

Here χ is the phase of the order parameter (or Δ) and is constant in the present case. If the above relation is substituted into Eq. (3.82) one obtains

$$J = (2\pi eRN_A(0)N_B(0))^{-1}\iint d\omega_1 d\omega_2 \tilde{\mathscr{D}}_A(\omega_1)\tilde{\mathscr{D}}_B(\omega_2)[f(\omega_1)-f(\omega_2)]$$
$$\times\left\{\mathrm{P}\frac{1}{\omega_2-\omega_1+eV}\sin(\chi_B-\chi_A-2eVt)\right.$$
$$\left.+\pi\delta(\omega_2-\omega_1+eV)\cos(\chi_B-\chi_A-2eVt)\right\}, \quad (3.84)$$

where P stands for the principal part. Note, from Eq. (3.82), that $\tilde{\mathscr{D}}_A\tilde{\mathscr{D}}_B \propto |\Delta_A||\Delta_B|$. Note also that J depends on the phase difference of the pair wave functions and hence oscillates with frequency $2eV$ in the presence of a potential difference. The second term in the curly brackets of Eq. (3.84) has the same structure as the quasi-particle term and is considered to be a contribution from pair tunnelling. Although the existence of this term has been confirmed experimentally, we do not consider it further here. The first term gives the well-known *Josephson effect*. The contribution of this term, denoted by J_{s1}, will be considered in the following.

Note first of all that if there exists a phase difference $\chi_B - \chi_A$ between two superconductors A and B, then J_{s1} is finite, even when $V = 0$. This corresponds to a supercurrent induced by a spatial variation of the order parameter phase in a bulk sample. The current here is carried by pairs keeping coherence while they tunnel. Let us write

$$J_{s1} = J_c \sin(\chi_B - \chi_A) \quad (3.85)$$

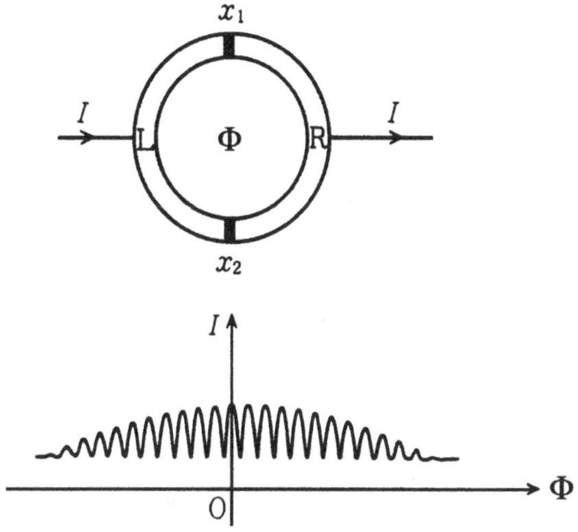

Fig. 3.10. The interference effect due to an external magnetic flux Φ piercing through the ring.

when $V = 0$. The energy of the tunnelling junction associated with this current is

$$E_{\mathrm{T}} = -J_{\mathrm{c}} \cos(\chi_{\mathrm{B}} - \chi_{\mathrm{A}}), \tag{3.86}$$

which takes its minimum when $\chi_{\mathrm{B}} = \chi_{\mathrm{A}}$. Suppose both sides of the junction are made of the same superconducting metal ($\Delta = \Delta_{\mathrm{A}} = \Delta_{\mathrm{B}}$). In the BCS model the above coefficient is given by

$$J_{\mathrm{c}} = \frac{\pi}{2}(eR)^{-1}\Delta \tanh \frac{\beta\Delta}{2}. \tag{3.87}$$

In the limit $T \ll \Delta_{\mathrm{A}}, \Delta_{\mathrm{B}}$, on the other hand, J_{c} becomes

$$J_{\mathrm{c}} = 2(eR)^{-1}\frac{\Delta_{\mathrm{A}}\Delta_{\mathrm{B}}}{\Delta_{\mathrm{A}}+\Delta_{\mathrm{B}}} K\left(\frac{|\Delta_{\mathrm{A}}-\Delta_{\mathrm{B}}|}{\Delta_{\mathrm{A}}+\Delta_{\mathrm{B}}}\right),$$

where K is a complete elliptic integral of the first kind.

Interference effect

Suppose there are two Josephson junctions in a loop of superconducting wire and a flux Φ pierces through the loop as shown in Fig. 3.10. If the effect of the current is negligible except at the junctions, one may invoke the same argument as that for the flux quantisation in Chapter 1 to find

$$\chi_{\mathrm{R}}(x_1) - \chi_{\mathrm{L}}(x_1) + \chi_{\mathrm{L}}(x_2) - \chi_{\mathrm{R}}(x_2) = 2e\oint \boldsymbol{A}\cdot \mathrm{d}\boldsymbol{l} = 2\pi\Phi/\phi_0.$$

If the two junctions have identical characteristics, the supercurrent flowing from the left side to the right side is given by

$$I = 2J_c \sin\chi \cos(\pi\Phi/\phi_0),$$

where $\chi = \chi_R(x_1) - \chi_L(x_1) + \pi\Phi/\phi_0$. The phase of the wave through the two paths changes and interferes as the magnetic flux Φ varies. The SQUID (superconducting quantum interference device) makes use of this principle to measure a magnetic flux. It can measure a very small magnetic field of the order of 10^{-7} G.

a.c. Josephson effect

It can be seen from Eq. (3.84) that an alternating current of frequency

$$\hbar\omega = 2eV_0 \tag{3.88}$$

appears when a potential difference V_0 is applied to a Josephson junction (we use the subscript 0 to denote an applied direct current voltage). This is called the *a.c. Josephson effect*. The voltage $V = 1 \times 10^{-6}$ V corresponds to $2\pi\omega = 483.6$ MHz according to Eq. (3.88). The frequency of an a.c. Josephson current determines the potential difference. As a method to measure this, let us consider the case in which an a.c. voltage of frequency ω_1 is applied together with a d.c. voltage V_0. The first term of Eq. (3.85) is written as

$$J_{s1} = J_c \sin\chi(t)$$

where $\chi(t)$ is the phase difference between A and B and it is assumed that V is not very large. Let us recall here that the scalar potential always appears with the time-derivative of the phase of the wave function due to the gauge invariance requirement. We have, in the present case, a pair made of two electrons, and hence the relevant combination is $2eV + \partial\chi/\partial t$ where we have put $\hbar = 1$. Since $V = 0$ implies χ =constant, the time-variation of χ should satisfy

$$\partial\chi/\partial t + 2eV = 0. \tag{3.89}$$

If $V = V_0$, this becomes $\chi = -2eV_0t$ + const. and reproduces Eq. (3.84). If the solution of Eq. (3.89) for $V = V_0 + v\cos\omega_1 t$ is substituted into the above expression for J_{s1}, one obtains

$$J(t) = J_c \sin[2eV_0t + (2ev/\omega_1) \times \sin\omega_1 t + \text{const.}].$$

The Fourier expansion of the above expression reveals the important fact

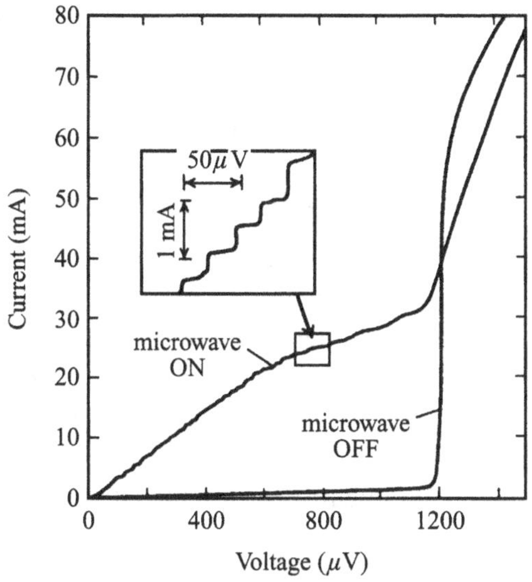

Fig. 3.11. The steps in the $I - V$ characteristic of an Sn–Sn tunnelling junction ($T = 1.2$ K) in the presence of a 10 GHz microwave. (From [G-11].)

that $J(t)$ has a d.c. component if

$$2eV_0 = n\omega_1 \quad (n \text{ is an integer})$$

is satisfied.

This effect is observed in an experiment in which the $I - V$ characteristic of a junction is measured while a microwave of frequency ω_1 is applied. It is seen from Fig. 3.11 that steps are observed at $V = n(\hbar\omega/2e)$, which are called the *Shapiro steps*.

A quantitative analysis of the characteristics requires a treatment in which the current due to quasi-particles is also considered. Since the microwave frequency measurement can be done with a relative error of the order of 10^{-12}, the ratio $2e/h = 4.8359767 \times 10^{14}$ Hz/V can be employed as a voltage standard thanks to this effect.

There are many topics concerning the Josephson effect which we have left untouched. See [C-4] and [D-1] for further information.

3.5 Mean-field theory in the presence of spatial variation

In the presence of an inhomogeneity due to an external field, boundaries or impurities, the energy gap also varies in space, which produces localised excitations and scattering of quasi-particle excitations. It is convenient to introduce real space formalism to treat this kind of problem. Let us write the Hamiltonian in the mean-field approximation as

$$\mathscr{H}_{\text{mf}} = \int \mathrm{d}\boldsymbol{x}\ \{\psi_\alpha^\dagger H_{(\text{b})}\psi_\alpha + \psi_\alpha^\dagger U(\boldsymbol{x})\psi_\alpha$$
$$+\Delta^*(\boldsymbol{x})\psi_\uparrow(\boldsymbol{x})\psi_\downarrow(\boldsymbol{x}) + \Delta(\boldsymbol{x})\psi_\downarrow^\dagger(\boldsymbol{x})\psi_\uparrow^\dagger(\boldsymbol{x})\}. \tag{3.90}$$

Here $H_{(\text{b})}$ is a one-particle Hamiltonian in which the band effect of the periodic potential is taken into account. This is given by $(1/2m^*)(\nabla/\mathrm{i} - e\boldsymbol{A}/c)^2 - \mu$ in the free electron model. The constant $\Delta^*\Psi$ has been dropped here. $U(\boldsymbol{x})$ is the potential due to impurities and boundaries, and ordinary corrections due to the Hartree–Fock approximation have been renormalised into the chemical potential μ. The attractive interaction is approximated by $-g\delta(\boldsymbol{x}-\boldsymbol{x}')$ as before and the mean field for the superconductivity is

$$\Delta(\boldsymbol{x}) = -g\langle\psi_\uparrow(\boldsymbol{x})\psi_\downarrow(\boldsymbol{x})\rangle. \tag{3.91}$$

The equations of motion for ψ and $\psi^\dagger$ in the Heisenberg picture are

$$\mathrm{i}\frac{\partial\psi_\uparrow}{\partial t} = [H_{(\text{b})} + U(\boldsymbol{x})]\psi_\uparrow - \Delta(\boldsymbol{x})\psi_\downarrow^\dagger$$
$$-\mathrm{i}\frac{\partial\psi_\downarrow^\dagger}{\partial t} = [H_{(\text{b})}^* + U(\boldsymbol{x})]\psi_\downarrow^\dagger - \Delta^*(\boldsymbol{x})\psi_\uparrow. \tag{3.92}$$

They are different from the ordinary Hartree–Fock equations in that $\psi_\uparrow$ and $\psi_\downarrow^\dagger$ couple to each other. We have assumed for simplicity that $U(\boldsymbol{x})$ is independent of spin. Corresponding to the Bogoliubov transformation (3.2), the fields ψ and $\psi^\dagger$ are expanded in terms of the creation and annihilation operators $\gamma_{n\uparrow}$ and $\gamma_{n\downarrow}^\dagger$ of the quasi-particle in the eigenstate n as

$$\begin{pmatrix} \psi_\uparrow \\ \psi_\downarrow^\dagger \end{pmatrix} = \sum_n \begin{pmatrix} u_n(\boldsymbol{x}) & v_n(\boldsymbol{x}) \\ -v_n{}^*(\boldsymbol{x}) & u_n{}^*(\boldsymbol{x}) \end{pmatrix} \begin{pmatrix} \gamma_{n\uparrow} \\ \gamma_{n\downarrow}^\dagger \end{pmatrix}. \tag{3.93}$$

Here $(u_n(\boldsymbol{x}), v_n(\boldsymbol{x}))$ is the eigenfunction and reduces to the plane wave state $(\mathrm{e}^{\mathrm{i}kx}u_k, \mathrm{e}^{\mathrm{i}kx}v_k)$ in the homogeneous case. If one substitutes the *Ansatz* $\gamma_{n\uparrow}, \gamma_{n\downarrow}^\dagger \propto \mathrm{e}^{-\mathrm{i}\varepsilon_n t}$ into Eq. (3.92) and sets the coefficient of $\gamma_{n\uparrow}$ (or $\gamma_{n\downarrow}^\dagger$) to 0, one obtains the Bogoliubov–de Gennes equation which determines $u_n(\boldsymbol{x})$ and $v_n(\boldsymbol{x})$,

$$\begin{pmatrix} H_{(\text{b})} + U(\boldsymbol{x}) - \varepsilon_n & \Delta(\boldsymbol{x}) \\ -\Delta^*(\boldsymbol{x}) & H_{(\text{b})}{}^* + U(\boldsymbol{x}) + \varepsilon_n \end{pmatrix} \begin{pmatrix} u_n(\boldsymbol{x}) \\ v_n{}^*(\boldsymbol{x}) \end{pmatrix} = 0. \tag{3.94}$$

It can be shown from Eq. (3.94) that if (u_n, v_n) is a solution with the eigenvalue ε_n, then (u_n^*, v_n^*) is a solution with the eigenvalue $-\varepsilon_n$. Thus we

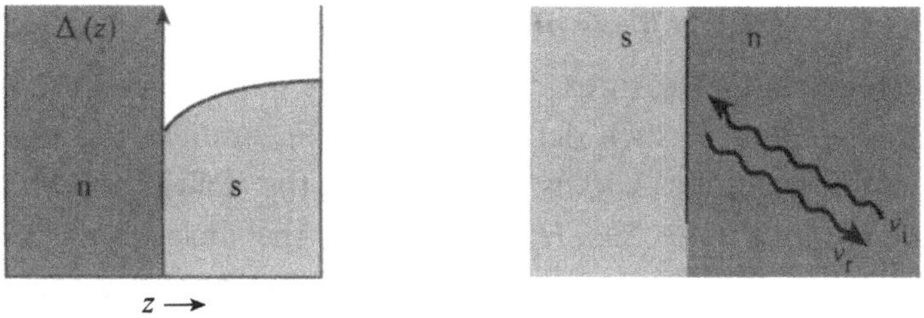

(a) (b)

Fig. 3.12. (a) The spatial variation of the energy gap when a superconductor is in contact with a normal metal. (b) Andreev scattering. The velocity vector of the excitation is reversed.

need to consider the quasi-particle excitations with a positive energy only. Note that eigenfunctions with different eigenvalues are orthogonal to each other. The gap $\Delta(\boldsymbol{x})$ is determined from Eqs. (3.91) and (3.93) as

$$\Delta(\boldsymbol{x}) = +g \sum_n u_n(\boldsymbol{x}) v_n(\boldsymbol{x})[1 - 2f(\varepsilon_n)] \tag{3.95}$$

and it is then left for us is to find the solutions (u_n, v_n) which satisfy both Eqs. (3.94) and (3.95) simultaneously. In the absence of a magnetic field, the assumption of uniformity readily leads to the results of Section 3.1.

The Bogoliubov–de Gennes equations (3.94) and (3.95) are employed to find the energy gap $\Delta(\boldsymbol{x})$ and the quasi-particle excitation in the presence of inhomogeneity induced by boundaries or an external magnetic field. For example, when a superconductor is in contact with a normal metal or a different superconductor, the gap $\Delta(\boldsymbol{x})$ varies in the vicinity of the contact, see Fig. 3.12(a). A quasi-particle scattering due to the spatial variation of Δ (the *Andreev scattering*) takes place in this case as shown in Fig. 3.12(b). Suppose a particle is incoming with momentum $\boldsymbol{k}$ from the normal side (n). On entering the superconducting side (s), the particle forms a pair with a particle with momentum $-\boldsymbol{k}$ and the pair becomes a part of the condensate. Accordingly a hole with momentum $-\boldsymbol{k}$ remains in the n side. The reflected wave in this case propagates in the opposite direction to the incoming wave and this should be compared with ordinary scattering where only the component perpendicular to the reflection plane changes sign. It is known that the magnetic flux quantum penetrates into a type 2 superconductor and there exist localised states in the vicinity of the flux. The Bogoliubov–de Gennes equation is used to analyse these states in Chapter 5. (See [C-1].)

3.5.1 Effects of defects and boundaries

One of the remarkable features of an ordinary superconductor is that T_c is almost independent of such defects as surfaces, boundaries, nonmagnetic impurities or dislocations. Let us assume that the defect potential $U(\boldsymbol{x})$ is independent of spin. One has $H^*_{(b)} = H_{(b)}$ in the absence of an external magnetic field ($A = 0$). The operator in Eq. (3.94) becomes real if $\Delta(\boldsymbol{x})$ is taken to be real (a constant phase $e^{i\chi}$ may be removed since $v(\boldsymbol{x})$ may be redefined to have the same phase). This implies that there exist real solutions (u_n, v_n) with degenerate eigenvalues $\varepsilon_{n\uparrow} = \varepsilon_{n\downarrow} = \varepsilon_n$. Therefore Eq. (3.95) justifies the assumption that $\Delta(\boldsymbol{x})$ is real. This can also be explained from the viewpoint of pairing. Although $U(\boldsymbol{x})$ is quite a complicated potential, in general, the eigenvalue equation

$$[H_{(b)} + U(\boldsymbol{x})]\phi_n(\boldsymbol{x}) = \xi_n\phi_n(\boldsymbol{x})$$

always has a set of one-particle wave functions $\{\phi_n\}$, which form a complete orthonormal basis. It is clear that if $\phi_{n\uparrow}$ is a solution with the eigenvalue ξ_n then $\phi^*_{n\downarrow}$ is also a solution with the same eigenvalue. Accordingly, the theory becomes identical to that for a uniform case if electrons in the states $\phi_{n\uparrow}$ and $\phi^*_{n\downarrow}$ are to form a pair. The pair wave function in the present case is $\langle\phi_{n\uparrow}(\boldsymbol{x})\phi^*_{n\downarrow}(\boldsymbol{x})\rangle$ and is real as mentioned above.

In a singlet s-wave superconductor, a one-electron state $\psi_{n\alpha}$, α being the spin, and its time-reversed state $R_t\psi_{n\alpha}$ generally form a pair. Accordingly, the pair is made of the set $(\psi_{n\uparrow}(\boldsymbol{x}), \psi^*_{n\downarrow}(\boldsymbol{x}))$ as above, since angular momentum changes sign under time-reversal. In the presence of impurities or boundaries, a pair is made of a one-particle state (and its time-reversal) in which scattering from these defects is already taken into account. Furthermore, if impurity atoms are distributed randomly so that one may assume $\Delta(\boldsymbol{x})$ is independent of $\boldsymbol{x}$, one obtains from Eqs. (3.94) and (3.95) the gap equation

$$\Delta = g\sum_n \frac{\Delta}{2\varepsilon_n}\tanh\frac{\beta\varepsilon_n}{2}, \qquad \varepsilon_n = \sqrt{\xi_n{}^2 + \Delta^2}, \tag{3.96}$$

which is identical to that for a uniform case. Therefore the gap Δ does not change, unless the density of states near the Fermi surface is modified by impurities. (It should be noted that impurities affect the scattering process if the interaction and the density of states are anisotropic near the Fermi surface.) Although the electron density changes, of course, within the screening distance r_s from the boundary or an impurity atom, this effect may be negligible since r_s is small.

An important example of cases without time-reversal symmetry is a magnetic impurity, which leads to depairing. A system carrying a supercurrent has been treated in Section 3.2, where it was shown that a pair wave function has a phase factor $e^{i\boldsymbol{q}\cdot\boldsymbol{x}}$ and is no longer real. Then the above argument is not applicable and even non-magnetic impurities lead to depairing. The Hamiltonian satisfies $H_{(b)} \neq H_{(b)}^*$ in the presence of an external magnetic field, which necessarily induces spatial variation of the phase. In general the superconducting state is then dramatically altered.

3.6 Gor'kov equations

The Green's function method, which is useful in many-body theory, is an indispensable tool in the theory of superconductivity. This is evident, for example, in the theory of electron–phonon interaction which will be treated in the next chapter. In the present section, we formulate the BCS theory in terms of the Green's functions, which will be helpful in the following discussions. (For the generalisation to imaginary time formalism and other topics, see [A-1] and [A-4] for example.)

The one-particle Green's function is defined by

$$G_{\alpha\beta}(\boldsymbol{x},t;\boldsymbol{x}',t') \equiv -\mathrm{i}\langle \mathrm{T}\psi_\alpha(\boldsymbol{x},t)\psi_\beta^\dagger(\boldsymbol{x}',t')\rangle. \tag{3.97}$$

Here $\psi_\alpha(\boldsymbol{x},t)$ is the electron field operator in the Heisenberg representation, T is the time-ordering operator (for $t' > t$, the average becomes $\mathrm{i}\langle\psi_\beta^\dagger(\boldsymbol{x}',t')\psi_\alpha(\boldsymbol{x},t)\rangle$) and $\langle\cdots\rangle$ denotes the statistical average. There exists a pair condensate in the superconducting state, where we restrict ourselves within the spin-singlet pairings. Then, in addition to G, one has to introduce

$$\begin{aligned} F(\boldsymbol{x},t;\boldsymbol{x}',t') &\equiv \langle \mathrm{T}\psi_\uparrow(\boldsymbol{x},t)\psi_\downarrow(\boldsymbol{x}',t')\rangle \\ F^\dagger(\boldsymbol{x},t;\boldsymbol{x}',t') &\equiv \langle \mathrm{T}\psi_\downarrow^\dagger(\boldsymbol{x},t)\psi_\uparrow^\dagger(\boldsymbol{x}',t')\rangle. \end{aligned} \tag{3.98}$$

It is convenient in a uniform system to introduce the Fourier transformed function such as

$$G_{\alpha\beta}(\boldsymbol{k},\omega) = -\mathrm{i}\int_{-\infty}^{\infty} \mathrm{d}t e^{\mathrm{i}\omega t}\langle \mathrm{T}a_{k\alpha}(t)a_{k\alpha}^\dagger(0)\rangle\delta_{\alpha\beta}.$$

Let us denote an eigenstate of the total system with the eigenvalue E_n by $|E_n\rangle$. The Green's function G is obtained from the definition of the statistical average as

$$\begin{aligned} \operatorname{Re} G(\boldsymbol{k},\omega) &= +\int_{-\infty}^{\infty} \mathrm{d}x\rho^{(G)}(\boldsymbol{k},\omega)\mathrm{P}\frac{1}{\omega - x} \\ \operatorname{Im} G(\boldsymbol{k},\omega) &= -\pi\rho^{(G)}(\boldsymbol{k},\omega)\tanh(\omega/2T). \end{aligned} \tag{3.99}$$

Here $\rho^{(G)}$ is just the spectral density defined by Eq. (3.73). Similarly the Green's function

$$F(\boldsymbol{k},\omega)=\int_{-\infty}^{\infty}\mathrm{d}t e^{\mathrm{i}\omega t}\langle \mathrm{T}a_{k\uparrow}(t)a_{-k\downarrow}(0)\rangle$$

can be written in terms of the spectral density $\rho^{(F)}(\boldsymbol{k},\omega)$ of Eq. (3.80).

The Green's function above is defined at an arbitrary temperature. At finite temperatures $T\neq 0$, however, finite temperature Green's functions employing imaginary time must be introduced in order to use perturbation theory based on Wick's theorem. Accordingly we restrict ourselves to the case $T=0$; we will be concerned with the ground state Green's functions only. The equations of motion that the Green's functions G and F satisfy are obtained readily from Eq. (3.92) as

$$\begin{aligned}&\left[\mathrm{i}\frac{\partial}{\partial t}-H_{(\mathrm{b})}-U(\boldsymbol{x})\right]G(\boldsymbol{x},t;\boldsymbol{x}',t')-\mathrm{i}\Delta(\boldsymbol{x},t)F^{\dagger}(\boldsymbol{x},t;\boldsymbol{x}',t')=\delta(\boldsymbol{x}-\boldsymbol{x}')\delta(t-t')\\&\left[\mathrm{i}\frac{\partial}{\partial t}+H_{(\mathrm{b})}{}^{*}+U(\boldsymbol{x})\right]F^{\dagger}(\boldsymbol{x},t;\boldsymbol{x}',t')+\mathrm{i}\Delta^{*}(\boldsymbol{x},t)G(\boldsymbol{x},t;\boldsymbol{x}',t')=0.\end{aligned} \tag{3.100}$$

The mean field $\Delta(\boldsymbol{x})$ is obtained from the definition of F as

$$\Delta(\boldsymbol{x},t)=gF(\boldsymbol{x},t_{+};\boldsymbol{x},t). \tag{3.101}$$

Here $t_{+}=t+\Delta t$ $(\Delta t>0)$ and $\Delta t\to 0$ is understood. The potential U contains the Hartree–Fock mean field. Equations (3.100) are called the *Gor'kov equations.*

Gor'kov equations are quite useful in analysing spatially nonuniform superconducting states in the presence of a magnetic field or impurities for example. We note *en passant* that the nonvanishing matrix elements appearing in the mean field are of the form $\langle N-2|\psi\psi|N\rangle$. Thus $\Delta(\boldsymbol{x},t)$ has a time dependent factor $e^{-\mathrm{i}2\mu t}$. Similarly $\Delta^{*}(\boldsymbol{x},t)$ is proportional to $e^{\mathrm{i}2\mu t}$. The a.c. Josephson effect in the presence of a potential difference, which we studied in Section 3.4.2, is related to this factor.

As a preliminary to the next chapter, we calculate the Green's functions G,F and $F^{\dagger}$ for a uniform case in the absence of an external field. It is also assumed that spin–orbit interaction is absent. Let us first introduce the following Nambu formalism to show that Feynman–Dyson perturbation theory is applicable to superconducting states. We first note that

$$\hat{a}_{k}=\begin{pmatrix}a_{k\uparrow}\\a^{\dagger}_{-k\downarrow}\end{pmatrix},\quad \hat{a}_{k}{}^{\dagger}=\begin{pmatrix}a^{\dagger}_{k\uparrow}, & a_{-k\downarrow}\end{pmatrix} \tag{3.102}$$

satisfy the fermionic commutation relation

$$\{\hat{a}_k(t), \hat{a}_k^\dagger(t)\} = \hat{1}.$$

From this observation, one defines the following 2×2 Green's function

$$\begin{aligned} \widehat{G}(\boldsymbol{k}, t) &\equiv -\mathrm{i}\langle \mathrm{T}\hat{a}_k(t)\hat{a}_k^\dagger(0)\rangle \\ &= \begin{pmatrix} G_{\uparrow\uparrow}(\boldsymbol{k}, t) & -\mathrm{i}F(\boldsymbol{k}, t) \\ -\mathrm{i}F^\dagger(\boldsymbol{k}, t) & -G_{\downarrow\downarrow}(-\boldsymbol{k}, -t) \end{pmatrix}. \end{aligned} \tag{3.103}$$

Now the Hamiltonian $H_{(\mathrm{b})}$ is written, with the help of Eq. (3.102), as

$$H_{(\mathrm{b})} = \sum_{k,\alpha} \xi_k a_{k\alpha}^\dagger a_{k\alpha} = \sum_k (\hat{a}_k^\dagger \xi_k \hat{\tau}_3 \hat{a}_k + \xi_k \hat{1}), \tag{3.104}$$

where τ_i's are the Pauli matrices and ξ_k is the conduction band energy including the Hartree–Fock energy. The free particle Green's function without interactions is given by

$$G_0{}^{-1}(\boldsymbol{k}, \omega) = (\omega\hat{1} - \xi_k \hat{\tau}_3). \tag{3.105}$$

The inter-particle interaction (2.17) is written in the form of a 2×2 matrix as

$$H_\mathrm{I} = -\frac{1}{2} \sum_{k,k',q} V_{k'k} \hat{a}_{k'+q}^\dagger \tau_3 a_{k+q} \hat{a}_{-k'}^\dagger \tau_3 a_{-k}$$

which can also be written as

$$H_\mathrm{I} = -\frac{1}{4} \sum V_{k'k} \hat{a}_{k'+q}^\dagger (\hat{\tau}_1 + \mathrm{i}\hat{\tau}_2) \hat{a}_{k'} \cdot \hat{a}_k^\dagger (\hat{\tau}_1 - \mathrm{i}\hat{\tau}_2) \hat{a}_{k+q}. \tag{3.106}$$

Therefore one may regard $\hat{a}_k^\dagger$ and $\hat{a}_k$ as two-component fermion operators and use Wick's theorem in the perturbation theory, if only one keeps in mind that the Green's functions and the interactions are of matrix form. Therefore one may employ the ordinary Feynman diagram technique.

In the mean-field approximation one obtains from Eq. (3.106) the Hamiltonian

$$H_\mathrm{I} = -\sum_k \hat{a}_k^\dagger (\Delta_k^{(1)} \hat{\tau}_1 + \Delta_k^{(2)} \hat{\tau}_2) \hat{a}_k \tag{3.107}$$

where we have put $q = 0$ and

$$\Delta_k^{(1,2)} = \sum_{k'} V_{k'k} \langle \hat{a}_{k'}^\dagger \tau_{1,2} \hat{a}_{k'} \rangle. \tag{3.108}$$

With these definitions, the mean-field Hamiltonian is written in a compact form as

$$\widehat{H}_{\rm mf} = \sum_k \hat{a}_k^\dagger(\xi_k\hat{\tau}_3 - \widehat{\Delta})\hat{a}_k \\ \widehat{\Delta} = \Delta_k^{(1)}\hat{\tau}_1 + \Delta_k^{(2)}\hat{\tau}_2, \tag{3.109}$$

from which one readily obtains the matrix Green's function

$$\widehat{G}(\boldsymbol{k},\omega) = \frac{1}{\omega^2 - \varepsilon_k{}^2}(\omega\hat{1} + \xi_k\hat{\tau}_3 - \widehat{\Delta}), \tag{3.110}$$

where $\varepsilon_k^2 = \xi_k^2 + |\Delta_k|^2$. This result may be derived, of course, from the Gor'kov equations (3.100) directly. The gap equation is just Eq. (3.108) and $\Delta_k^{(1)}$ and $\Delta_k^{(2)}$ are determined by similar equations. $\Delta_k^{(1)}$ and $\mathrm{i}\Delta_k^{(2)}$ are real numbers and the χ defined by $|\Delta_k|\mathrm{e}^{\mathrm{i}\chi} = \Delta_k^{(1)} + \Delta_k^{(2)}$ is the phase of the condensate. It is noted that $\Delta_k^{(1),(2)}$ are complex numbers for a phonon-mediated attractive force.

Two final remarks are in order. The Gor'kov equations and the Bogoliubov–de Gennes equation are equivalent formalisms in the weak-coupling approximation, where there is no time delay in the interaction. In cases in which averages over the impurity positions are involved, however, the Gor'kov equations must be employed since what are to be averaged are not the wave functions but the Green's functions (the correlation functions).

It is exceedingly difficult to solve the Gor'kov equations in the presence of a magnetic field or in the analysis of the proximity effect, which appears when a superconducting metal is in contact with a normal metal. The crudest approximation which is useful to treat these systems is the Ginzburg–Landau theory, which is applicable when the scale of the spatial variation L of the system satisfies $L \gg \xi_0$, see Chapter 5. The semiclassical approximation (G. Eilenberger, 1968) is a useful method when $\xi_0 \gtrsim L \gg k_{\rm F}^{-1}$, see [C-2] for details. It is assumed there that the amplitude of all the quasi-particle momenta $\boldsymbol{k}$ is $k_{\rm F}$ and the energy associated with the spatial variation is approximated by $v_{\rm F}\hat{\boldsymbol{k}}\cdot\nabla$. We note that if the pair has an angular momentum other than the s-wave, the mean field Δ_k depends on the direction of $\boldsymbol{k}$ (see Chapter 6) and accordingly one has to introduce the $\boldsymbol{x}$-dependent field $\Delta_k(\boldsymbol{x})$. Thus the semiclassical approximation must be introduced from the outset.

4

Superconductivity due to electron–phonon interaction

In most superconducting materials including metals such as Al, Pb and Nb and metallic compounds such as Nb_3Sn, the electron pair is induced by the phonon-mediated interaction between electrons. This interaction is simplified in the BCS theory and is replaced by a direct attractive force between electrons. Here we discuss a theory in which electron–phonon interaction is considered from the beginning. In so doing an approximation to the Coulomb repulsion between electrons is also taken into account. In band theory, which serves as the foundation of the theory of electrons in solid state physics, one considers the Schrödinger equation with periodic potential of ions resting on the crystal lattice together with the mean-field potential that incorporates to some extent the Coulomb interaction between electrons. As a result, one-electron states expressed by Bloch functions are obtained as the eigenfunctions. These states are specified by the band index, the wave number $\boldsymbol{k}$ in the first Brillouin zone and the spin α, and will be denoted simply by $\boldsymbol{k}, \alpha$. Furthermore, the spin index α is dropped in many cases since we will not treat a system with magnetism unless so stated. It is assumed in the following that the energy eigenvalue ξ_k is already known through the band theory.

4.1 Electron–phonon system

The motion of ions is treated as a lattice vibration. The normal mode is specified by the wave number $\boldsymbol{q}$ and the polarisation σ, and will be denoted simply by q. Let $b_q^\dagger$ and b_q be the creation and the annihilation operators of a phonon which has been obtained by quantising a mode with frequency ω_q and wave number q. Then the displacement of the ions associated with a lattice vibration with mode q is given by

$$u_q = (2NM\omega_q)^{-1/2}(b_q + b_{-q}^\dagger). \tag{4.1}$$

Here we have assumed for simplicity that there are N ions of a single kind with mass M. There is electron scattering in the presence of a lattice vibration since electrons move in a potential which is not perfectly periodic. The electron–phonon interaction that causes scattering is

$$\mathscr{H}_{\text{e-ph}} = \sum_{k,k',\alpha} M_{k'k} u_q a^\dagger_{k'\alpha} a_{k\alpha} \tag{4.2}$$

to the first order in the displacement of ions. The matrix element in the above Hamiltonian is given by

$$M_{k'k} = -\int \mathrm{d}\boldsymbol{x} \sum_l \boldsymbol{n}_q \cdot \nabla U(\boldsymbol{x} - \boldsymbol{R}_l)\phi^*_{k'}(\boldsymbol{x})\phi_k(\boldsymbol{x}) \exp \mathrm{i}[\boldsymbol{q}\cdot\boldsymbol{R}_l + (\boldsymbol{k}-\boldsymbol{k}')\cdot\boldsymbol{x}]. \tag{4.3}$$

Here U is the potential due to ions, $\boldsymbol{R}_l$ the position of the lth ion in the absence of lattice vibration, $\phi_k(\boldsymbol{x})\mathrm{e}^{\mathrm{i}\boldsymbol{k}\cdot\boldsymbol{x}}$ the Bloch function and $\boldsymbol{n}_q$ the unit polarisation vector of lattice vibration with wave number vector $\boldsymbol{q}$. The wave number $\boldsymbol{q}$ is determined by $\boldsymbol{k}' - \boldsymbol{k} = \boldsymbol{q} + \boldsymbol{G}$, $\boldsymbol{G}$ being the reciprocal lattice vector. In the following, however, the Umklapp processes in which $\boldsymbol{G} \neq 0$ are not explicitly written. Summarising the above results, one writes the total Hamiltonian of a system of electrons and a lattice as

$$\begin{aligned}\mathscr{H} &= \sum_k \xi_k a^\dagger_k a_k + \sum_q \omega_q \left(b^\dagger_q b_q + \frac{1}{2}\right) \\ &\quad + \sum_{k,q} \frac{1}{\sqrt{2NM\omega_q}} M_{k+q,k}(b_q + b^\dagger_{-q}) a^\dagger_{k+q} a_k + \mathscr{H}_{\text{Coulomb}}.\end{aligned} \tag{4.4}$$

The final term is to be considered as a part of the Coulomb interaction between electrons, which has not been taken into account in the band energy ξ_k.

The method of Green's functions is useful to analyse this system. The electron Green's functions have already been defined in Section 3.6. The phonon Green's function is defined by

$$\begin{aligned}D(q,t) &\equiv -\mathrm{i}\langle \mathrm{T} u_q(t) u_{-q}(0)\rangle \\ &= -\frac{\mathrm{i}}{2NM\omega_q}\langle \mathrm{T}\{b_q(t)b^\dagger_q(0) + b^\dagger_{-q}(t)b_{-q}(0)\}\rangle\end{aligned} \tag{4.5}$$

since the phonon operators always appear in the combination u_q of Eq. (4.1). Corresponding to Eq. (3.99), it is convenient to express the Fourier transform $D(\boldsymbol{q}, \omega)$ in terms of the phonon spectral density $\rho_{\rm ph}$ as

$$D(\boldsymbol{q}, \omega) = \int_0^\infty dx \rho_{\rm ph}(\boldsymbol{q}, x) \left\{ \frac{1}{\omega - x + i\delta} - \frac{1}{\omega + x - i\delta} \right\} \tag{4.6}$$

$$\rho_{\rm ph}(\boldsymbol{q}, x) = \frac{1}{2NM\omega_q} \sum_n |\langle 0|b_q|n\rangle|^2 \delta(x - E_n), \tag{4.7}$$

where $|0\rangle$ and $|n\rangle$ are the ground state and an excited state with energy E_n respectively. Here use has been made of the relation $\omega_{-q} = \omega_q$ and only the form at $T = 0$ has been given. When there are no interactions of phonons with each other or with electrons, we have the free phonon spectral density

$$\rho_{\rm ph}{}^{(0)}(\boldsymbol{q}, x) = (2NM\omega_q)^{-1} \delta(x - \omega_q). \tag{4.8}$$

There are three physically important effects of electron–phonon interaction:

1. Electron scattering by phonons. This contributes to the lifetime τ and manifests itself in transport phenomena such as the electrical resistivity.
2. Renormalisation of the electron energy. In particular this leads to an increase of the effective mass m^*, which can be distinguished from the effect of the periodic potential. This may be observed through specific heat data.
3. It leads to the electron–electron interaction. This was first proposed by Frölich as the origin of superconductivity, see [C-2] for example.

As a preliminary to superconductivity, which is the main theme of the present chapter, in the next section we study the effects of electron–phonon interaction in the normal state, that is, the effects 1 and 2 mentioned above.

4.2 Electron–phonon interaction in the normal state

Important information on the system, such as the density of the excited states, is obtained once the Green's functions are known. Let G_0 and D_0 be the electron and the phonon Green's functions, respectively, in the absence of any interaction. In Fig. 4.1 they are represented by thin solid and wavy lines respectively, while those with interaction are thick solid and wavy lines in Fig. 4.2. The interaction is denoted by a vertex which emits or absorbs

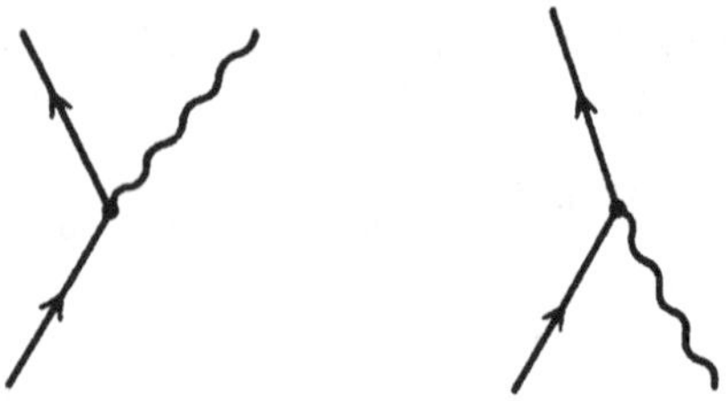

Fig. 4.1. The electron–phonon interaction.

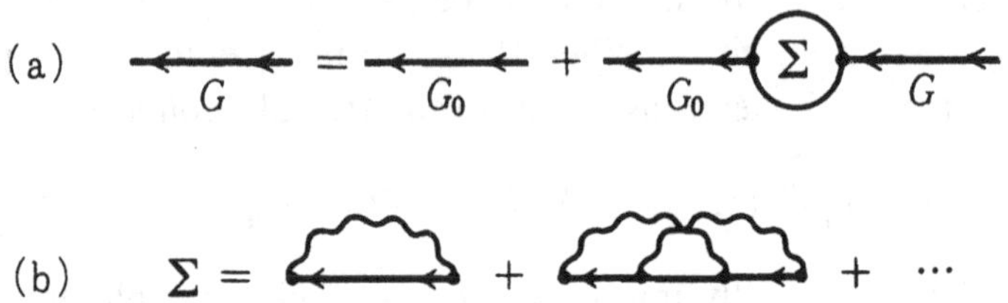

Fig. 4.2. The Feynman diagrams for (a) the electron Green's function and (b) the self-energy.

a phonon. It is necessary to evaluate the self-energies $\Sigma(\boldsymbol{k}, \omega)$ and $\Pi(\boldsymbol{q}, \omega)$ defined by the Feynman–Dyson equations

$$\begin{aligned} G(\boldsymbol{k}, \omega) &= G_0(\boldsymbol{k}, \omega)[1 + \Sigma(\boldsymbol{k}, \omega) G(\boldsymbol{k}, \omega)] \\ &= [G_0{}^{-1}(\boldsymbol{k}, \omega) - \Sigma(\boldsymbol{k}, \omega)]^{-1} \end{aligned} \tag{4.9}$$

$$D(\boldsymbol{q}, \omega) = [D_0{}^{-1}(\boldsymbol{q}, \omega) - \Pi(\boldsymbol{q}, \omega)]^{-1} \tag{4.10}$$

to obtain the full Green's functions G and D, see Figs. 4.2 and 4.3.

The self-energies $\Sigma(\boldsymbol{k}, \omega)$ and $\Pi(\boldsymbol{q}, \omega)$ are given by infinite summations of the Feynman diagrams and are shown in Figs. 4.2(b) and 4.3(b), respectively.

Corresponding to the direct interaction $V_{kk'}$ between electrons, one has the quantity $|M_{kk'}|^2 D(\boldsymbol{k} - \boldsymbol{k}', \omega - \omega')$, which is one of the fundamental building blocks in the perturbative expansion. It is common to introduce the combination

$$g(\boldsymbol{k}, \boldsymbol{k}', x) \equiv N(0)|M_{kk'}|^2 \rho_{\text{ph}}(\boldsymbol{k} - \boldsymbol{k}', x) \tag{4.11}$$

since $|M_{kk'}|^2$, containing all information of the individual system, is always multiplied by the spectral density $\rho_{\text{ph}}(\boldsymbol{q}, \omega)$ (we note that the symbol $\alpha^2 F$ is often used, instead of g, in literature). It is assumed here that the system is of unit volume. It will be shown below that $g(\boldsymbol{k}, \boldsymbol{k}', x)$ is a dimensionless quantity and will prove to be useful in characterising the strength of the electron–phonon interaction. It is important to realise that the phonon-mediated

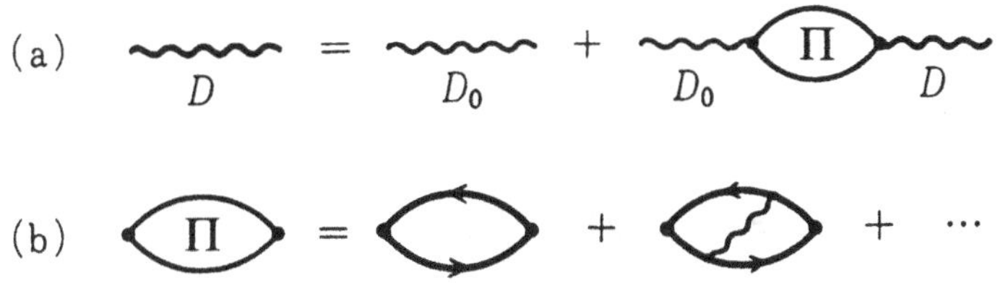

Fig. 4.3. The Feynman diagrams for (a) the phonon Green's function and (b) the self-energy.

Fig. 4.4. The processes that are considered in the electron self-energy calculation.

interaction depends on the variable x which corresponds to the frequency, in contrast with the direct interaction. In other words, the interaction is time-dependent since the action takes time to propagate.

The main purpose of this section is to calculate the self-energy Σ of an electron. As for a phonon, the spectrum ω_q, at least, is known experimentally through neutron diffraction and other methods, as we will see later. Hence, supposing the effect of the interaction is to renormalise phonon frequency to the actually observed value ω_q, one can neglect the phonon self-energy $\Pi(\boldsymbol{q}, \omega)$ in Eq. (4.10). Note however that such a scheme is not allowed if phonons are 'softened' by an interaction with electrons. Let us assume that the electron self-energy Σ is approximately given by the first term of Fig. 4.2(b). That the other terms are negligible is justified by Migdal's theorem, which we will touch upon later. Even in this approximation, all the diagrams in Fig. 4.4 are taken into account. The corresponding expression is

$$\Sigma(\boldsymbol{k}, \omega) = -\frac{1}{N(0)} \sum_{\boldsymbol{k}'} \int_0^\infty \mathrm{d}x \; g(\boldsymbol{k}, \boldsymbol{k}', x) \times \int_{-\infty}^{\infty} \frac{\mathrm{d}\omega'}{2\pi\mathrm{i}} \left(\frac{1}{\omega - \omega' - x + \mathrm{i}\delta} - \frac{1}{\omega - \omega' + x - \mathrm{i}\delta} \right) G(\boldsymbol{k}', \omega'). \tag{4.12}$$

This is an integral equation since G in the right hand side contains Σ. Our purpose is to obtain $\Sigma(\boldsymbol{k}, \omega)$ for $|\boldsymbol{k}| = k_\mathrm{F}$ and $|\omega|$ of the order of the Debye frequency $\omega_\mathrm{D} \ll \varepsilon_\mathrm{F}$. Accordingly most contributions in the ω'-integration come from the region $\omega' \sim \omega_\mathrm{D}$ since ω, $x \sim \omega_\mathrm{D}$. Then it follows that most contributions in the $\boldsymbol{k}'$-summation come from those $\boldsymbol{k}'$ with $|\boldsymbol{k}'| \sim k_\mathrm{F}$. Making use of these facts, one puts $|\boldsymbol{k}| = |\boldsymbol{k}'| = k_\mathrm{F}$ and replaces the $\boldsymbol{k}'$-summation by

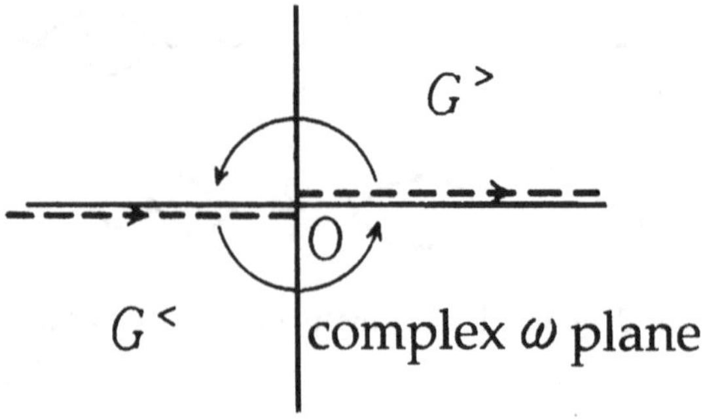

Fig. 4.5. The modification of the integration contour in the self-energy calculation.

$\sum_{k'} \simeq N(0) \int \frac{d\Omega_{k'}}{4\pi} \int d\xi_{k'}$. However the ω'-integration should be carried out first.

In the limit $T \to 0$, Eq. (3.99) becomes

$$G(\boldsymbol{k}, \omega) = \int_{-\infty}^{\infty} dx \frac{\rho^{(G)}(\boldsymbol{k}, x)}{\omega - x + i(\omega/|\omega|)\delta}. \tag{4.13}$$

Thus the function G is defined separately for $\omega > 0$ and $\omega < 0$ and, if ω is continued to the complex plane, these functions are analytic in the upper half plane and the lower half plane, respectively. By making use of this fact, one can change the contour of ω'-integration to the path shown in Fig. 4.5, which runs slightly below the negative real axis and slightly above the positive real axis. As a result, the ω'-integration in Eq. (4.12) is rewritten as

$$\int_{-\infty}^{\infty} \frac{d\omega'}{2\pi i} \left(\frac{\theta(\omega')}{\omega - \omega' - x + i\delta} + \frac{\theta(-\omega')}{\omega - \omega' + x - i\delta} \right) 2i\, \mathrm{Im}G(\boldsymbol{k}', \omega').$$

According to Eq. (4.13), $\mathrm{Im}G(\boldsymbol{k}', \omega')$ is just the spectral density. If written in this form, one easily finds that the ξ'-integration converges in the limit $|\xi'| \to \infty$. The self-energy $\Sigma(\boldsymbol{k}', \omega')$ is considered to be independent of ξ' since $\omega_D \ll \varepsilon_F$ (see Migdal's theorem below). Then

$$\int_{-\infty}^{\infty} d\xi' G = - \int_{-\infty}^{\infty} d\xi' \left[\xi' - \omega' + \Sigma(\omega')\right]^{-1}$$

can simply be obtained by residue calculus. In doing so we must keep in mind that the signature of $\mathrm{Im}\Sigma(\omega)$ is $\omega/|\omega|$. Finally we reach the following result,

$$\begin{aligned} \Sigma(\omega) &= \int_0^{\infty} dx \bar{g}(x) \int_0^{\infty} dz \left\{ \frac{1}{\omega - z - x + i\delta} + \frac{1}{\omega + z + x - i\delta} \right\} \\ &= \int_0^{\infty} dx \bar{g}(x) \left\{ \ln \left| \frac{\omega - x}{\omega + x} \right| - i\pi(\omega/|\omega|)\theta(|\omega| - x) \right\}, \end{aligned} \tag{4.14}$$

where

$$\bar{g}(x) \equiv \int \frac{d\Omega_{k'}}{4\pi} g(\hat{\boldsymbol{k}}, \hat{\boldsymbol{k}}', x). \tag{4.15}$$

Here the system is assumed isotropic. An important point is that the ξ'-integration can be carried out in Eq. (4.14) owing to the approximation that is applicable when $\omega_D \ll \varepsilon_F$. Accordingly the result is identical with that obtained by second order perturbation theory.

4.2.1 *Effective mass and lifetime*

When the quasi-particle description is applicable, the Green's function near the Fermi surface is approximated by

$$G^{-1}(\boldsymbol{k}, \omega + i\delta) \simeq (1 + \lambda_k)\omega - (\xi_k + \Sigma^{(r)}(\boldsymbol{k}, 0)) - i\Sigma^{(i)}(\boldsymbol{k}, \omega)$$

$$\lambda_k \simeq -\left.\frac{\partial \Sigma(\boldsymbol{k}, \omega)}{\partial \omega}\right|_{\omega=0} = 2\int_0^\infty dx \frac{\bar{g}(x)}{x}. \tag{4.16}$$

Here the superscripts (r) and (i) stand for the real and imaginary parts respectively. Since the real part $\Sigma^{(r)}(\boldsymbol{k}, 0)$ vanishes within the approximation employed in Eq. (4.14), the peak of the spectral density ImG appears at $\omega = \xi_k/(1 + \lambda_k)$. In the isotropic case ($\lambda_k = \lambda$) the quasi-particle mass becomes

$$m \to (1 + \lambda)m.$$

The variable λ is, therefore, called the *mass enhancement parameter*. The quasi-particle lifetime τ_k is given by

$$1/2\tau_k \simeq -(1 + \lambda_k)^{-1}\Sigma^{(i)}(\boldsymbol{k}, (1 + \lambda_k)^{-1}\xi_k). \tag{4.17}$$

One has to know the coupling constant g to evaluate Σ explicitly. The phonon spectrum and the matrix element $M_{kk'}$ may be evaluated by the pseudo-potential method for a relatively simple metal. We will not persue that here and, instead, use phonons in the so-called *jellium model* and calculate g. Electrons move in the following screened Coulomb potential of ions

$$U(q) = 4\pi e^2/q^2\varepsilon(q) = 4\pi e^2/(q^2 + k_s{}^2) \tag{4.18}$$

$$k_s{}^2 = 6\pi n e^2/\varepsilon_F.$$

We have assumed that the ion is univalent and employed the Thomas–Fermi permeability. Then Eq. (4.3) becomes

$$|M_{k+q,k}|^2 = [4\pi n e^2 q/(q^2 + k_s{}^2)]^2. \tag{4.19}$$

In this model, phonons consist only of longitudinal mode and their dispersion relation is given by

$$\omega_q{}^2 = 4\pi ne^2q^2/M(q^2 + k_s{}^2).$$

When $q \ll k_s$, one has $\omega_q = v_s q$ with the sound velocity $v_s = \sqrt{m/3M} \cdot v_F$, which is reduced from v_F by a factor proportional to $\sim \sqrt{m/M}$. If, on the other hand, $q \geq k_s$ is satisfied, ω_q approaches the ion plasma frequency $\Omega_q^2 = 4\pi ne^2/M$. From Eqs. (4.8), (4.11) and (4.19), one obtains

$$g(\boldsymbol{k} + \boldsymbol{q}, \boldsymbol{k}, x) = \frac{1}{4}\frac{k_s{}^2}{q^2 + k_s{}^2}\omega_q\delta(x - \omega_q).$$

From this expression one can get a rough idea of g. In the present approximation, one has $q \simeq k_F\sqrt{1-\mu}$ with $\mu = \boldsymbol{k}\cdot\boldsymbol{k}'/k_F^2$. Then the average over the Fermi surface is equivalent to that over μ and one finds

$$\bar{g}(x) = \frac{k_s{}^2}{4k_F{}^2}\frac{x^2}{\Omega_p{}^2 - x^2}. \tag{4.20}$$

If this expression is used in Eq. (4.14), one obtains

$$\Sigma^{(i)}(\omega) = -\pi\frac{k_s{}^2}{4k_F{}^2}\left\{\frac{\Omega_p}{2}\ln\frac{\Omega_p + \omega}{\Omega_p - \omega} - \omega\right\}. \tag{4.21}$$

Therefore, when $|\omega| \ll \Omega_p$, the imaginary part is given by $\Sigma^{(i)} \sim -\pi\omega^3/6k_F^2v_s^2$ and one finds that the description in terms of particle-like excitations is applicable in the vicinity of the Fermi surface and the lifetime at low temperatures ($T \ll k_F v_s$) is given by $\tau^{-1} \sim T^3$, regarding ω as T. Furthermore, from Eqs. (4.16) and (4.20),

$$\lambda = \frac{k_s{}^2}{4k_F{}^2}\ln\left(1 + \frac{4k_F{}^2}{k_s{}^2}\right), \tag{4.22}$$

where the phonon wave number is assumed to satisfy $q < 2k_F$. The ratio k_s/k_F is given in terms of the ratio of the Bohr radius $a_0 = 1/me^2$ and the mean electron separation r_0 as

$$\frac{k_s{}^2}{4k_F{}^2} = \left(\frac{4}{9}\right)^{1/3}\pi^{-4/3}\frac{r_0}{a_0} \sim \frac{r_0}{6a_0}.$$

The ratio r_0/a_0 is of the order of unity for ordinary metals.

According to band theory, the electron specific heat at low temperatures is proportional to the density of states $2N(0) = m_b k_F/\pi^2$, where m_b is the band

Table 4.1. *Observed values of m_{th} and the mass enhancement parameters λ.*

Element	m_{th}/m	m_{b}/m	$m_{\text{th}}/m_{\text{b}}$	$1+\lambda$
Li	2.22	1.54	1.44	1.41
Na	1.24	1.01	1.22	1.16
K	1.21	1.07	1.13	1.13
Rb	1.37	1.19	1.15	1.16
Cs	1.80	1.53	1.18	1.15
Al	1.49	1.05	1.42	1.44

Remark: m_{b} and λ are obtained from the pseudo potential theory, see [D-2].

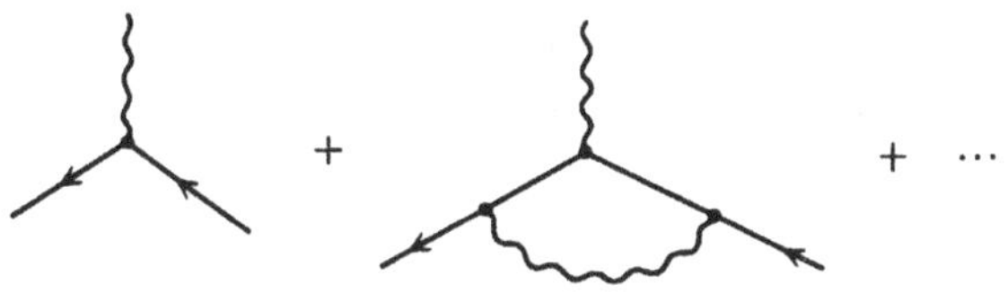

Fig. 4.6. Vertex corrections.

mass. Let m_{th} be the electron mass determined from the observed specific heat. Then we expect these two masses to satisfy

$$m_{\text{th}}/m_{\text{b}} = 1 + \lambda. \tag{4.23}$$

Table 4.1 shows that for simple metals the theoretical estimates obtained by the pseudo potential theory agree very well with the observed values.

4.2.2 Migdal's theorem

When we evaluated the self-energy Σ, we neglected the higher order diagrams in Fig. 4.3(b). These higher order processes are the so-called vertex corrections and Migdal's theorem states that these corrections are proportional to the powers of $\sqrt{m/M}$ and hence small. Origin of the lowest order correction is illustrated in Fig. 4.6 and is given by

$$\overline{|M|^2} \sum_{q'} D(q')G(k-q')G(k+q-q')$$

where k denotes $\boldsymbol{k}$ and ω collectively. We have replaced the matrix element $M_{k+q,k}$ by its average since it is not a rapidly varying quantity. The dominant contribution to the self-energy Σ comes from the region where both k and

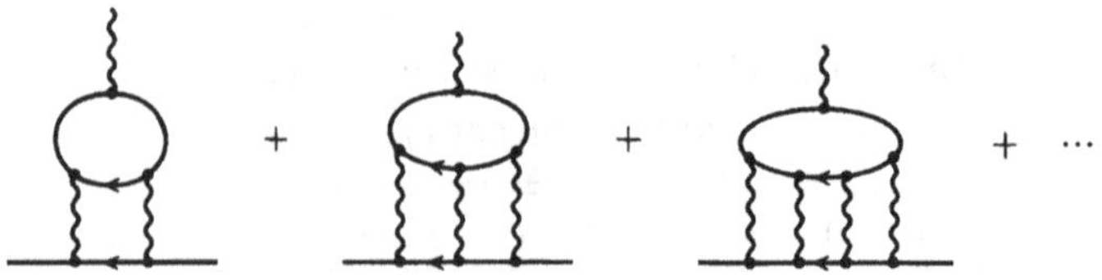

Fig. 4.7. Vertex corrections and pair formation.

$k+q$ are near the Fermi surface. Since most phonons have wave number q of the order of k_F, either $\boldsymbol{k}-\boldsymbol{q}'$ or $\boldsymbol{k}+\boldsymbol{q}-\boldsymbol{q}'$ is far from the Fermi surface and accordingly $\xi \sim \varepsilon_F$. Then one of the G's becomes ε_F^{-1} and the above integral is of the order of $\omega_D/\varepsilon_F \sim \sqrt{m/M}$. There is an exceptional case in which all of $\boldsymbol{k}, \boldsymbol{k}+\boldsymbol{q}$ and $\boldsymbol{k}-\boldsymbol{q}'$ are close to the Fermi surface (in the sense that $|\xi_k| < \omega_D$) and the fourth vector $\boldsymbol{k}+\boldsymbol{q}-\boldsymbol{q}'$ is also close to the Fermi surface. However the contribution from this process is a small quantity proportional to $N(0)\omega_D/n \sim v_s/v_F \sim \sqrt{m/M}$.

Migdal's theorem is not applicable to a system that can be unstable against the formation of a charge density wave or a spin density wave due to nesting. This is the case for a system with a quasi-one-dimensional Fermi surface. We also note that the channel leading to superconductivity is made of particle–particle ladder diagrams and hence the vertex correction is as illustrated by the diagrams in Fig. 4.7. Although each diagram is proportional to a power of $\sqrt{m/M}$ according to Migdal's theorem, the infinite sum of the series diverges. Accordingly one cannot apply perturbation theory to these processes.

4.3 Eliashberg equation

The strong coupling theory of superconductivity due to electron–phonon interaction starts with the Eliashberg equation, which is an extension of the Gor'kov equations. This extension amounts to replacing the electron–electron interaction $V_{kk'}$ by $g(\boldsymbol{k},\boldsymbol{k}',x)$. It should be noted, however, that a phonon propagates with a finite velocity. Consequently there is a *retardation* in the interaction and the mean field itself depends on the excitation energy as will be shown below. It should also be noted that although the Hartree–Fock term in the Gor'kov equations may be absorbed by simply renormalising ξ_k, one has to deal directly with the ordinary self-energy due to the many-body effect in the present theory.

In the 2×2 matrix form introduced in Section 3.6, the self-energy parts appearing in the Green's function $\widehat{G}$ are given by the Feynman–Dyson equation, which is an extension of Eq. (4.9) in the previous section, as

$$\widehat{\Sigma}(\boldsymbol{k},\omega) = \widehat{G}_0^{-1}(\boldsymbol{k},\omega) - \widehat{G}^{-1}(\boldsymbol{k},\omega). \tag{4.24}$$

Here $\widehat{G}(\boldsymbol{k},\omega)$ is the Fourier transform of $\widehat{G}$ defined by Eq. (3.103) while $\widehat{G}_0(\boldsymbol{k},\omega)$ is obtained similarly from Eq. (3.105). The electron–phonon interaction, that is, the third term in Eq. (4.4) is obtained by replacing $a^\dagger_{k+q}a_k$ by a matrix $\hat{a}^\dagger_{k+q}\hat{\tau}_3\hat{a}_k$ in this formalism, see Eq. (3.106). Therefore evaluating $\widehat{\Sigma}$ with the Feynman diagrams in the previous section amounts to applying the same approximation to the superconducting state. Then one obtains, corresponding to Eq. (4.12), the self-energy

$$\widehat{\Sigma}(\boldsymbol{k},\omega) = -\frac{1}{N(0)}\sum_{k'}\int\frac{\mathrm{d}\omega'}{2\pi\mathrm{i}}\int_0^\infty \mathrm{d}x g(\boldsymbol{k},\boldsymbol{k}',x)$$
$$\times\left(\frac{1}{\omega-\omega'-x+\mathrm{i}\delta} - \frac{1}{\omega-\omega'+x-\mathrm{i}\delta}\right)\hat{\tau}_3\widehat{G}(\boldsymbol{k}',\omega')\hat{\tau}_3. \tag{4.25}$$

Let us consider the general relation between $\widehat{G}$ and $\widehat{\Sigma}$ before we start explicit calculations. We again note that there is no magnetism in the system under consideration and hence G is independent of the spin and the system is invariant under spatial reflection, that is, $\widehat{G}(-\boldsymbol{k},\omega) = \widehat{G}(\boldsymbol{k},\omega)$. Let us write, therefore, $G_{\uparrow\uparrow}(\omega) = G_{\downarrow\downarrow}(\omega) = G(\omega)$ (we will omit the variable $\boldsymbol{k}$ in the following unless necessary). From the definition of G,

$$G_{11}(\omega) = G(\omega), \qquad G_{22}(\omega) = -G(-\omega). \tag{4.26}$$

Let us rewrite Eq. (4.24) as $(\widehat{G}_0^{-1} - \widehat{\Sigma})\widehat{G} = \widehat{1}$. This equation is equivalent to the set of four equations for the Green's functions defined by Eq. (3.103). One notes from Eq. (4.26), however, that there are only three independent Green's functions. If we write Eq. (4.24) for each component we find that the components $\Sigma_{\alpha\beta}$ of $\widehat{\Sigma}$ satisfy the relations

$$\begin{aligned}\Sigma_{22}(-\omega) &= -\Sigma_{11}(\omega)\\ \Sigma_{12}(\omega)F^\dagger(\omega) &= \Sigma_{21}(-\omega)F(-\omega).\end{aligned} \tag{4.27}$$

The components of $\widehat{G}$ are

$$\begin{aligned}G(\omega) &= [-G_0{}^{-1}(-\omega) + \Sigma_{11}(-\omega)]/d(\omega)\\ F^\dagger(\omega) &= \mathrm{i}\Sigma_{21}(\omega)/d(\omega), \qquad F(\omega) = \mathrm{i}\Sigma_{12}(\omega)/d(-\omega),\end{aligned} \tag{4.28}$$

Fig. 4.8. Two self-energies in a superconducting state.

where

$$d(\omega) = [-G_0{}^{-1}(-\omega) + \Sigma_{11}(-\omega)][G_0{}^{-1}(\omega) - \Sigma_{11}(\omega)] - \Sigma_{12}(\omega)\Sigma_{21}(\omega) \\ = d(-\omega). \tag{4.29}$$

If one substitutes Eq. (4.28) into Eq. (4.25), one finds that the off-diagonal components Σ_{12} and Σ_{21} of $\hat{\Sigma}$, which correspond to the energy gap function Δ_k, satisfy the same equation. Therefore the remark made for Δ_k at the end of the previous chapter equally applies to the off-diagonal components (the components Σ_{12} and Σ_{21} correspond to $\Delta_k^{(1)} + \Delta_k^{(2)}$ and $\Delta_k^{(1)} - \Delta_k^{(2)}$, respectively). Namely $e^{-i\chi}\Sigma_{21}(\omega) = e^{i\chi}\Sigma_{12}(\omega)$. The phase χ is associated with 'spontaneous symmetry breaking'. It can be shown that this equality is true in general and independent of the approximation employed in deriving Eq. (4.25). Accordingly one needs to consider $\Sigma_{12}(\omega)$ only, which is now an ω-dependent function unlike Δ in the BCS theory.

Considering these facts, let us define

$$\omega - \frac{1}{2}(\Sigma(\boldsymbol{k},\omega) - \Sigma(\boldsymbol{k},-\omega)) \equiv \omega Z(\boldsymbol{k},\omega) \\ \Sigma_1(\boldsymbol{k},\omega)/Z(\boldsymbol{k},\omega) \equiv \Delta(\boldsymbol{k},\omega), \tag{4.30}$$

where we have put $\Sigma_{11}(\omega) = -\Sigma_{22}(-\omega) = \Sigma$ and $\Sigma_{12} = \Sigma_1$. The self-energies Σ and Σ_1 correspond to the diagrams shown in Fig. 4.8. If one assumes that $\Sigma(\boldsymbol{k},0)$ is renormalised into ξ_k, one may neglect $\Sigma(\boldsymbol{k},\omega) + \Sigma(\boldsymbol{k},-\omega)$ for such values of ω that are used in the following. In terms of these quantities the solutions for a uniform system can be written as

$$\left.\begin{matrix} G \\ F^\dagger \end{matrix}\right\} = [\omega^2 Z^2 - \xi_k{}^2 - Z^2\Delta^2]^{-1} \times \begin{pmatrix} \omega Z + \xi_k \\ +iZ\Delta \end{pmatrix}. \tag{4.31}$$

It now remains to solve the nonlinear simultaneous equations obtained by substituting Eq. (4.31) into Eq. (4.25). An approximate calculation is possible since the dominant contribution in Eq. (4.25) comes from the region $|\boldsymbol{k}| \sim k_F$, in the same way as the normal state.

Let us rewrite the integrand of the ω-integral in terms of $\mathrm{Im}G$ and $\mathrm{Im}F^\dagger$

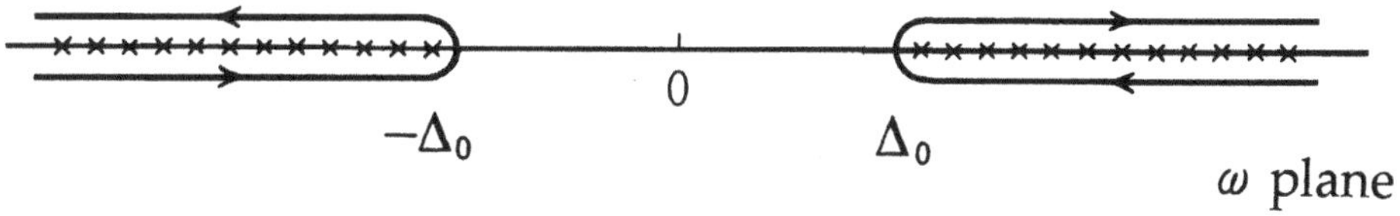

Fig. 4.9. The integration contour to evaluate Σ and Σ_1.

(that is, the difference of the Green's function in the upper and the lower half-planes along the cut) instead of G and $F^\dagger$, just as we did when we derived Eq. (4.14). In doing so, special note must be taken of the facts that (i) G and $F^\dagger$ satisfy the analytic property of ordinary Green's functions, namely they are analytic in the upper (lower) half-plane for $\omega > 0$ ($\omega < 0$) and (ii) cuts along the real line, that is the lines on which the denominators of G and $F^\dagger$ vanish, are given by $|\omega| > \Delta_0$ as shown in Fig. 4.9. The meaning of Δ_0 and Eq. (4.31) lead to the definition

$$\Delta_0 = \Delta(\Delta_0). \tag{4.32}$$

Thus the integration over $\xi_{k'}$ is carried out first. Then g is replaced by $\bar{g}(x)$ after being averaged over $\boldsymbol{k}$ as was done in Section 4.2. If the system is isotropic, Z and Δ become functions of ω only and Eq. (4.25) becomes

$$[1 - Z(\omega)]\omega = \int_{\Delta_0}^{\infty} \mathrm{d}z\, \mathrm{Re}\frac{z}{\sqrt{z^2 - \Delta^2(z)}}\lambda^{(+)}(\omega, z), \tag{4.33}$$

$$Z(\omega)\Delta(\omega) = \int_{\Delta_0}^{\infty} \mathrm{d}z\, \mathrm{Re}\frac{\Delta(z)}{\sqrt{z^2 - \Delta^2(z)}}\lambda^{(-)}(\omega, z), \tag{4.34}$$

where

$$\lambda^{(\pm)}(\omega, z) = \int_0^{\infty} \mathrm{d}x \bar{g}(x)\left(\frac{1}{\omega + z + x - \mathrm{i}\delta} \pm \frac{1}{\omega - z - x + \mathrm{i}\delta}\right). \tag{4.35}$$

To understand Eqs. (4.33)–(4.35), let us find a solution of these equations when all the phonons have a common frequency ω_E (Einstein phonons). Since we are interested in the gap equation $\Delta(\omega)$, we put $\Delta = 0$ in Eq. (4.33) and employ the normal state solution $Z(\omega) \simeq 1 + \lambda$, which is obtained with an assumption $\omega \ll \omega_\mathrm{E}$. The interaction for the Einstein phonons is given by

$$\bar{g}(x) = \frac{1}{2}\lambda\omega_\mathrm{E}\delta(x - \omega_\mathrm{E}). \tag{4.36}$$

If this expression is substituted into Eq. (4.16), one finds that the mass

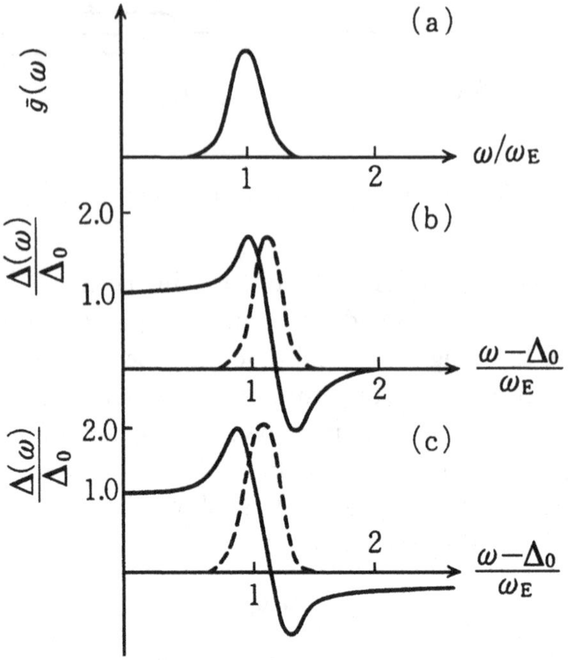

Fig. 4.10. (a) The phonon spectrum with a single phonon peak. This spectrum has been used to evaluate the solution $\Delta(\omega)$ whose real (imaginary) part is shown by the solid line (the broken line) in (b). (c) is the solution with the Coulomb interaction taken into account. (D.J. Scalapino: [C-2] vol. I.)

enhancement parameter is just given by this λ. If Eq. (4.36) is substituted into Eqs. (4.34) and (4.35), one obtains the integral equation

$$\Delta(\omega) = -\frac{\lambda}{1+\lambda}\omega_E \int_{\Delta_0}^{\infty} dz \, \mathrm{Re}\left(\frac{\Delta(z)}{\sqrt{z^2 - \Delta^2(z)}}\right) \frac{\omega_E + z}{(\omega)^2 - (\omega_E + z - i\delta)^2}. \quad (4.37)$$

Figure 4.10 shows numerically the phonon spectrum close to the Einstein phonon and the gap function obtained from the spectrum. Some important qualitative features are found from Eq. (4.37):

1. If one takes $\omega = \Delta_0$, one obtains an equation which determines the gap edge Δ_0. This equation has a nonvanishing solution Δ_0.
2. The real part of $\Delta(\omega)$ is positive (negative) for $\omega < \omega_E$ ($\omega > \omega_E$).
3. The imaginary part $\mathrm{Im}\Delta(\omega)$ appears for $\omega > \omega_E + \Delta_0$. At finite temperatures, however, $\mathrm{Im}\Delta(\omega)$ is finite even for $\omega < \omega_E + \Delta_0$.

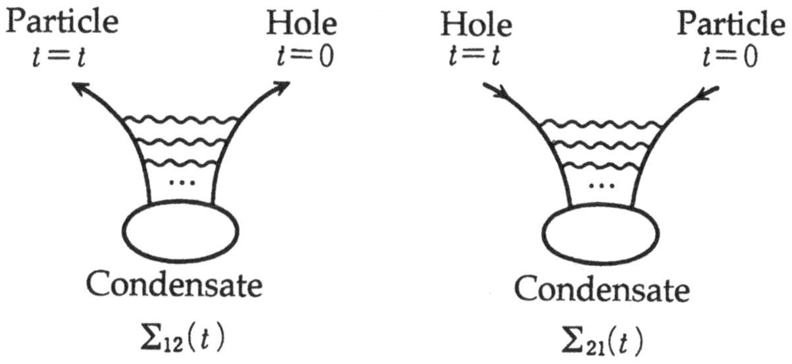

Fig. 4.11. The processes represented by Σ_{12} and Σ_{21}.

4.3.1 *Weak coupling*

The dominant contribution to the integral in Eq. (4.37) comes from the domain $z < \omega_E$ if $\lambda \ll 1$ namely $\Delta_0 \ll \omega_E$. Therefore we can approximate Eq. (4.37) by taking an appropriate upper limit $\omega_c \simeq \omega_E$, as

$$\Delta_0 \simeq \lambda \int_{\Delta_0}^{\omega_c} dz \frac{\Delta_0}{\sqrt{z^2 - \Delta_0{}^2}},$$

from which we obtain the BCS gap equation. We thus obtain $\Delta(\omega) = \text{const.} = \Delta_0$, $\omega < \omega_c$. This is the *weak coupling limit* where we have justified the BCS theory.

Before we close this section, let us comment on the ω-dependence of the gap function. Σ_{12} and Σ_{21} are the probability amplitudes for the processes shown in Fig. 4.11. Since it takes a finite time for a phonon to propagate, the amplitudes Σ_{12} and Σ_{21} are functions of t and hence complex functions of its Fourier transform ω. It is physically expected that these amplitudes are identical since the amplitude for a hole at $t = 0$ remaining a hole at later time $t = t$ is the same as that for a particle at $t = 0$ remaining a particle at $t = t$. The self-energy Σ_1 and hence the imaginary part of $\Delta(\omega)$ leads to the finite lifetime of a pair state due to phonon emissions (and scattering of thermally excited phonons at finite temperatures). The phase associated with symmetry breaking, namely the phase of the condensate can be understood as the phase of Δ at the gap edge, which has been taken as a real value of Δ_0 above. Even if Δ at the gap edge is chosen real, the function $\Delta(\omega)$ for $\omega \neq \Delta_0$ is in general complex.

Let us consider the present problem as that of a two-body bound state as mentioned in the beginning of Chapter 2. Since the interaction due to the exchange of phonons (pions for the nuclear force) involves retardation in

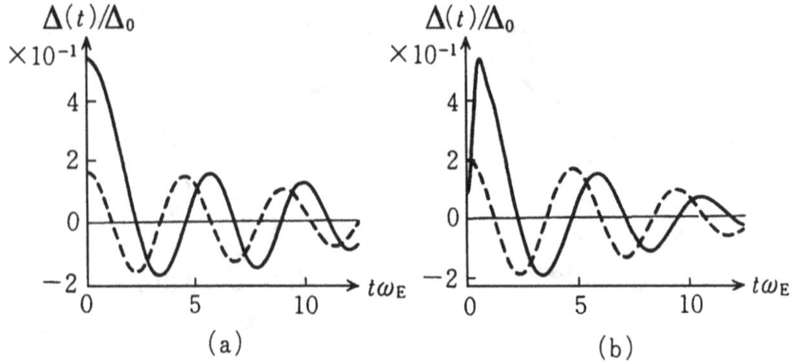

Fig. 4.12. The Fourier transforms of $\Delta(\omega)$. The solid (broken) line depicts the real (imaginary) part. Only the electron–phonon interaction is considered in (a) while the Coulomb interaction is also included in (b).

time, the two-body wave function must be thought of as a function of $\boldsymbol{x}_1, \boldsymbol{x}_2$ and t_1, t_2, or equivalently, $(1/2)(t_1 + t_2)$ and $t_1 - t_2$. If the wave function is Fourier transformed, it becomes $C(\boldsymbol{k}, \omega, \boldsymbol{q}, \Omega)$ instead of $C(\boldsymbol{k}, \boldsymbol{q})$ as in Section 2.1. Since the momentum and the energy are conserved if there are only two particles, those states with different $\boldsymbol{q}$ and Ω decouple in the equation of motion. The ω-dependence appears, however, in the equation corresponding to Eq. (2.8). We can regard this ω-dependence as the temporal extension of the wave function (see Fig. 4.12). It will play an important rôle in the effect of Coulomb repulsion studied in the next section.

4.4 Coulomb interaction

Needless to say, the repulsive Coulomb interaction V_c between electrons must be considered at the same time as the electron–phonon interaction. Unlike the phonon-mediated interaction, the ω-dependence of V_c appears when ω reaches the plasma frequency ω_p. Here V_c is assumed to have no retardation and hence be independent of ω, since $\omega_p > \omega_F$ in many cases. We say that the repulsive force V_c is *diluted* over the range $0 < |\omega| < \omega_p$, in contrast with the phonon-mediated attractive force which is localised at $|\omega| \sim \omega_D$. We will take advantage of this fact to treat the problem approximately.

First of all, contributions to Σ and Σ_1 are restricted to the lowest order diagrams in V_c (Fig. 4.13), which are formally the same as those for electron–phonon interaction. This amounts to keeping the Fock terms only. In this approximation, and dropping the summation and integration symbols, one simply adds $\Sigma^{(c)} = V_c \circ G$ and $\Sigma_1^{(c)} = V_c \circ F^\dagger$ to those terms due to electron–

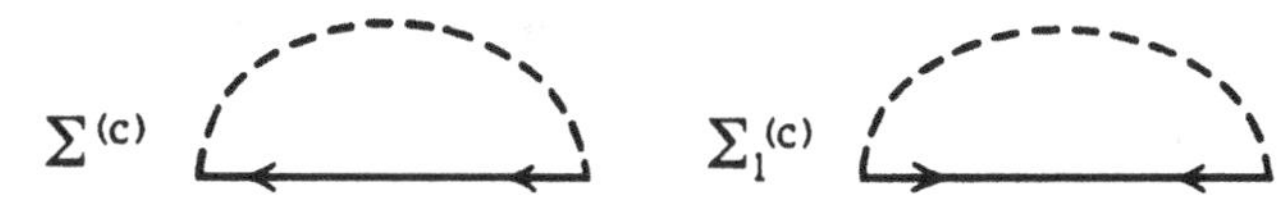

Fig. 4.13. The self-energies due to Coulomb interaction (dotted lines).

phonon interaction to obtain

$$\Sigma = \Sigma^{(ph)} + \Sigma^{(c)}, \qquad \Sigma_1 = \Sigma_1^{(ph)} + \Sigma_1^{(c)}. \tag{4.38}$$

Let us first consider $\Sigma^{(c)}$. The contribution $V_c \circ G_n$ to the self-energy in the normal state is expected to have a weak ω-dependence due to the reason mentioned above. Accordingly one can assume that this contribution is taken into account by renormalising ξ_k. Thus one only needs to consider $\Sigma^{(c)\prime} = V_c \circ (G - G_n)$. It should be noted, however, that G and G_n differ only in the vicinity of the Fermi surface while the contribution to V_c comes from a much broader area in momentum space. Therefore one may drop $\Sigma^{(c)\prime}$ and, as a result, one needs to consider only

$$\Sigma_1^{(c)}(\boldsymbol{k}) = -\sum_{k'} \int \frac{d\omega'}{2\pi i} V_c(\boldsymbol{k}, \boldsymbol{k}') F^\dagger(\boldsymbol{k}', \omega') \equiv V_c \circ F^\dagger. \tag{4.39}$$

In the above equation, the ω'-integration extends over a broad domain $|\omega'| \sim \omega_p \gg \omega_D$. Let us introduce a frequency ω_c, which is several times as large as ω_D, and divide the ω'-integration domain into I ($|\omega'| < \omega_c$) and II ($|\omega'| > \omega_c$). Since $|\Sigma_1^{(ph)}| \ll |\omega'|$ in region II, $F^\dagger$ is approximately given by

$$\begin{aligned} F^\dagger(\boldsymbol{k}', \omega') &\simeq K(\boldsymbol{k}', \omega') \Sigma_1^{(c)}(\boldsymbol{k}') \\ K(\boldsymbol{k}', \omega') &\equiv G_0(\boldsymbol{k}', \omega') G_0(-\boldsymbol{k}', -\omega') = -(\omega'^2 - \xi_{k'}{}^2)^{-1}. \end{aligned} \tag{4.40}$$

Note that $\Sigma_1^{(ph)}$ is also negligible in the numerator. By making use of this fact, one may introduce an effective interaction U in which the contribution from region II is incorporated and rewrite Eq. (4.39) into an integral equation over region I. For this purpose, let us consider U which satisfies an equation expressed by the diagram Fig. 4.14, which is of the same form as the T-matrix.

Here the intermediate states are restricted within region II, where $|\omega'| > \omega_c$. The diagram is written, in a symbolic form, as

$$U = V_c + V_c \circ K_{II} \circ U. \tag{4.41}$$

If this equation is compared with the following equations, obtained from

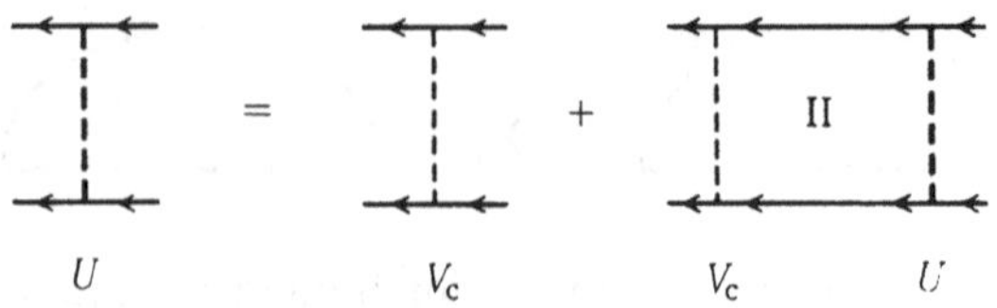

Fig. 4.14. The equation which determines the effective Coulomb interaction.

Eqs. (4.39) and (4.40),

$$\Sigma_1^{(c)} = V_c \circ F_I{}^\dagger + V_c \circ F_{II}{}^\dagger = V_c \circ F_I{}^\dagger + V_c \circ K_{II} \circ \Sigma_1^{(c)}$$

it is found that $UF_I^\dagger$ and $\Sigma_1^{(c)}$ satisfy the same equation. Therefore one may put

$$\Sigma_1^{(c)} = U \circ F_I{}^\dagger. \tag{4.42}$$

One may anticipate this relation by comparing Fig. 4.13 with Fig. 4.14 for the interaction U. Thus once a U that satisfies Eq. (4.41) is obtained, we can determine $\Sigma_1^{(c)}$ by solving the (nonlinear) integral equation (4.42), which contains an integration over region I only.

If a screened Coulomb interaction is assumed, the potential V_c is approximately constant over the region under consideration and hence U determined by Eq. (4.42) is also a constant. Then, taking the analyticity of K into account, only the following integration need be carried out,

$$N(0)\int_{|\omega'|>\omega_c} \frac{d\omega'}{2\pi i}\int_{-\varepsilon_b}^{\varepsilon_b} d\xi' K = N(0)\int_{\omega_c}^{\varepsilon_b} d\omega' \frac{1}{\omega'}.$$

Here the upper bound $\varepsilon_b \gg \omega_c$ of the $|\xi'|$-integration is considered to be of the order of the band width. The commonly used symbols $\mu \equiv N(0)V_c$ and $\mu^* \equiv N(0)U$ are related by

$$\mu^* = \frac{\mu}{1 + \mu \ln \dfrac{\varepsilon_b}{\omega_c}}. \tag{4.43}$$

If this relation is substituted into Eq. (4.42), one finally finds that the effect of the Coulomb interaction reduces to the replacement of Eq. (4.34) by

$$Z(\omega)\Delta(\omega) = \int_{\Delta_0}^{\omega_c} dz \,\mathrm{Re}\left(\frac{\Delta(z)}{\sqrt{z^2 - \Delta^2(z)}}\right)\{\lambda^{(-)}(\omega, z) - \mu^*\}. \tag{4.44}$$

The upper bound of the integral in the phonon part is replaced by ω_c, which is justified if ω_c is taken several times as large as ω_D.

Let us note that since V_c is nonvanishing for a wide range of frequencies, μ is weakened as in Eq. (4.43) and secondly, that μ^* is finite when λ in

Eq. (4.44) is peaked at $|z| \sim \omega_D$. The effectiveness of the phonon-mediated attraction can be understood from these reasons. The solution $\Delta(\omega)$ when μ^* is considered is also given in Fig. 4.10. One also finds that, when $\Delta(\omega)$ is Fourier transformed, the amplitude of $\Delta(t)$ is suppressed near $t = 0$ as is shown in Fig. 4.12.

Note, however, that Migdal's theorem is not applicable to the present problem since there is no retardation in the Coulomb interaction and hence it operates over a wide range of frequencies. Accordingly, it should be kept in mind that the approximate Σ obtained above has no quantitative meaning. We will discuss the size of the Coulomb interaction parameter μ^* in the next section.

4.5 Strong-coupling effects and critical temperature

4.5.1 Tunnelling effects

The ω-dependence of the energy gap Δ is directly observed by the tunnelling effect. As was mentioned in Section 3.4, the density of states $\mathscr{D}(\omega)$ can be obtained from the tunnelling characteristics between a superconductor and a normal metal at low enough temperatures, see Eq. (3.79). Using the Green's function (4.31), the density of states $\mathscr{D}(\omega)$ is obtained, in the present case, as

$$\mathscr{D}(\omega) = \frac{1}{2\pi \mathrm{i}} \sum_{k} \{G(\boldsymbol{k}, \omega + \mathrm{i}\delta) - G(\boldsymbol{k}, \omega - \mathrm{i}\delta)\} \tag{4.45}$$

and by simply putting $\Delta = \Delta(\omega)$. The result is

$$\mathscr{D}(\omega) = N(0)\mathrm{Re}\frac{|\omega|}{\sqrt{\omega^2 - \Delta^2(\omega)}}. \tag{4.46}$$

Note that Δ^2 in the square root is not $|\Delta|^2$. The density of states in BCS theory is obtained by the substitutions $\Delta = \Delta_0$ ($\Delta_0 = \Delta(\Delta_0)$) for $|\omega| < \omega_D$ and $\Delta = 0$ for $|\omega| > \omega_D$ in the above expression. The strong coupling effect is measured as a deviation from this BCS result. Figure 4.15 is an example of such a measurement for lead. The deviation is 10% at most even for such a strong-coupling metal.

The density of states (4.46) may be obtained, in principle, from the Eliashberg equation once $g(\boldsymbol{k}, \boldsymbol{k}', \omega)$, defined by Eq. (4.11), or its average $\bar{g}(\omega)$ and the Coulomb interaction parameter μ^* are known. However it is difficult, in practice, to evaluate these quantities theoretically. McMillan and Rowell [C-2], conversely, made an attempt to determine $\bar{g}(\omega)$ and μ^* numerically so that the measured tunnelling characteristics and Δ_0 might be obtained from the Eliashberg equation. They obtained $\mu^* = 0.12$ and $\bar{g}(\omega)$ shown in

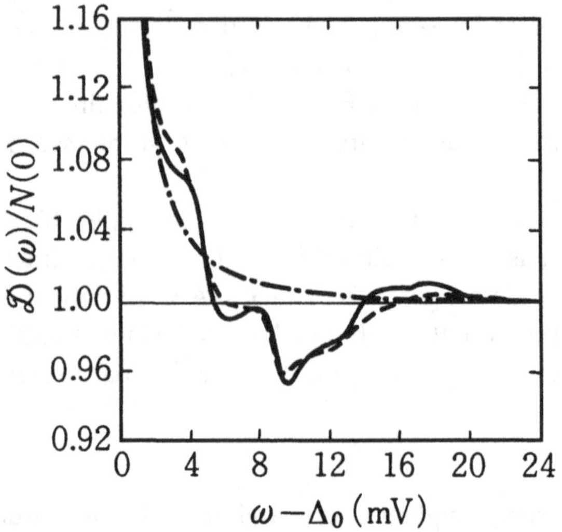

Fig. 4.15. The energy dependence of the density of states of lead in a superconducting state. The broken line represents the data obtained from the tunnelling effect while the BCS weak-coupling theory yields the dot-and-broken line. The strong-coupling result is shown by a solid line. (J.R. Schrieffer, *et al.*: [C-3].)

Fig. 4.16(a) for Pb. It is interesting to compare $\bar{g}(\omega)$ of Fig. 4.16(a) and the phonon density of states Fig. 4.16(b) obtained by neutron diffraction experiments. The renormalisation function $Z(\omega)$ and the amplitude $\Delta(\omega)$ of Pb in the same calculation are shown in Fig. 4.17(a) and (b). The peak of the phonon density of states (van Hove singularity) manifests itself in the structure of $\Delta(\omega)$. Since phonon emission is dominant in the vicinity of the peak, the imaginary part of Δ increases there.

4.5.2 Transition temperature

To study finite temperature properties such as the transition temperature T_c, the temperature Green's function is employed. In actual numerical computations, moreover, it is easy to solve equations with the discrete frequency (Matsubara frequency) $\omega_n = (2n+1)\pi T$, n being an integer. Here we only quote the equation which determines T_c,

$$\Delta(n) = \sum_{n'} \frac{1}{(2n'+1)} \left\{ \lambda(n-n') - \mu^* - \delta_{nn'} \sum_{n''} \lambda(n'-n'') \frac{n''}{|n''|} \cdot \frac{n'}{|n'|} \right\} \Delta(n'). \tag{4.47}$$

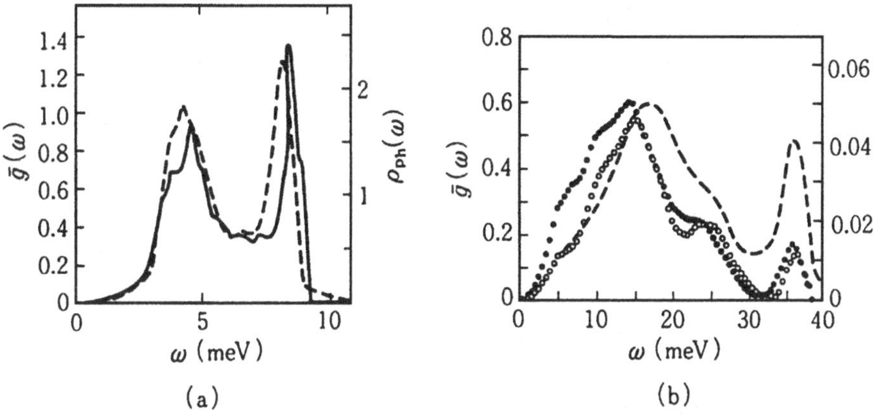

Fig. 4.16. (a) $\bar{g}(\omega)$ for Pb. The dotted line shows the phonon density of states obtained by neutron diffraction experimentz. (W.L. McMillan and J.M. Rowell: [C-2] vol. I.) (b) $\bar{g}(\omega)$ of Nb_3Al obtained by the tunnelling effect. • is for a sample with $T_c = 16.4$ K while ∘ is for $T_c = 14.0$ K. The dotted line denotes the phonon spectral density obtained by neutron scattering experiments. (From [G-12].)

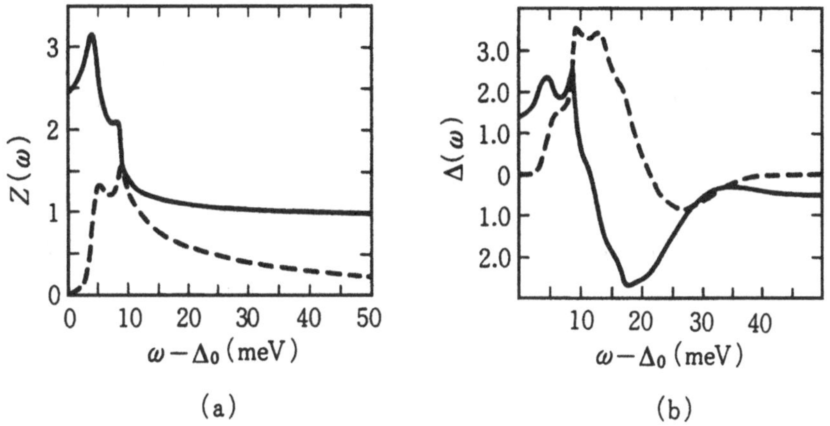

Fig. 4.17. (a) The real (solid line) and the imaginary (broken line) parts of the renormalisation function $Z(\omega)$ of Pb in the superconducting state, and (b) the real (solid line) and the imaginary (broken line) parts of the amplitude $\Delta(\omega)$ of Pb. (W.L. McMillan and J.M. Rowell: [C-2] vol. I.)

The function $\lambda(n - n')$ above corresponds to $\lambda(\omega, z)$ in Eq. (4.35). The above equation is obtained by taking $\Delta = 0$ in the denominator of the finite temperature version of Eqs. (4.33) and (4.34) and then eliminating $Z(\omega)$.

To find a convenient expression for T_c, we put

$$\lambda(n - n') = \lambda\theta(\omega_D - |\omega_n|)\theta(\omega_D - |\omega_{n'}|)$$

in the first term and $\lambda(n) = \lambda\theta(\omega_D - |\omega_n|)$ in the second term of Eq. (4.47) so that this equation becomes 'separable'. Here λ is the mass enhancement parameter. Then if $\Delta(n) = \Delta \cdot \theta(\omega_D - |\omega_{n'}|)$ is substituted into Eq. (4.47), one obtains

$$\frac{1+\lambda}{\lambda - \mu^*} = \psi\left(\frac{\omega_D}{2\pi T_c} + 1\right) - \psi\left(\frac{1}{2}\right),$$

where ψ is the di-Gamma function. From the above equation we finally find $T_c = 1.13\omega_D \exp\{-(1+\lambda)/(\lambda - \mu^*)\}$. We quote, *en passant*, McMillan's formula

$$k_B T_c = 0.83\omega_{\ln} \exp\left\{\frac{-1.04(1+\lambda)}{\lambda - \mu^*(1+0.62\lambda)}\right\}, \tag{4.48}$$

which has been employed most frequently in the discussion of T_c. Here

$$\omega_{\ln} = \exp\langle \ln \omega \rangle, \qquad \langle \ln \omega \rangle = \frac{2}{\lambda} \int_0^\infty \mathrm{d}\omega \bar{g}(\omega) \frac{\ln \omega}{\omega}.$$

There are other improved expressions for T_c, see [E-3].

4.5.3 Isotope effects

The first experimental evidence that electron–phonon interaction is responsible for superconductivity in ordinary metals was derived from the isotope effect, first observed in Sn and Hg. It may be assumed that the phonon frequency ω_q, and hence the Debye frequency ω_D or the averaged frequency $\omega_{\ln}$ is proportional to $M^{-1/2}$. It is seen from Eqs. (4.14) and (4.16) that λ is independent of M. Therefore only $\omega_{\ln}$ and μ^* containing ω_c depend on M in the expression for T_c. The isotope effect is measured by the index α defined by

$$\alpha \equiv -\frac{\mathrm{d}(\ln T_c)}{\mathrm{d}(\ln M)}. \tag{4.49}$$

One finds from Eqs. (4.48) and (4.43) that

$$\alpha = \frac{1}{2}\left\{1 - \frac{1.04(1+\lambda)(1+0.62\lambda)(\mu^*)^2}{[\lambda - \mu^*(1+0.62\lambda)]^2}\right\}. \tag{4.50}$$

One obtains the BCS value $\alpha = 1/2$ if μ^* is independent of M. One also notices that α is never greater than $1/2$.

4.5.4 Strong-coupling effects at finite temperatures

Thermal excitations tend to weaken the mean field as already mentioned. They also manifest themselves in a different form in the strong-coupling theory. To see this new effect, let us generalise Eq. (4.37), which determines $\Delta(\omega)$ in the Einstein model, to finite temperature cases. The result is

$$\Delta(\omega) = \frac{\lambda}{1+\lambda}\omega_{\mathrm{E}} \int_0^\infty \mathrm{d}z \, \mathrm{Re}\left(\frac{\Delta(z)}{\sqrt{z^2-\Delta^2(z)}}\right)$$
$$\times \frac{1}{2}\left\{\left[\tanh\frac{z}{2} + f(z) + n(\omega_{\mathrm{E}})\right]\left(\frac{1}{\omega+z+\omega_{\mathrm{E}}-\mathrm{i}\delta} - \frac{1}{\omega-z-\omega_{\mathrm{E}}+\mathrm{i}\delta}\right)\right.$$
$$\left. -[f(z)+n(\omega_{\mathrm{E}})]\left(\frac{1}{\omega-z+\omega_{\mathrm{E}}-\mathrm{i}\delta} - \frac{1}{\omega+z-\omega_{\mathrm{E}}+\mathrm{i}\delta}\right)\right\}, \quad (4.51)$$

where $n(x) = [\exp(x)-1]^{-1}$ is the Bose distribution function. The above equation reduces to the BCS expression in the weak-coupling limit. This equation also shows that $\Delta(\omega)$ is comparatively more suppressed by the finite temperature effect in the strong-coupling theory. Accordingly the ratio $2\Delta_0/T_{\mathrm{c}}$ becomes relatively large. Furthermore $\mathrm{Im}\omega$ remains finite at $T \neq 0$, even when $\omega < \omega_{\mathrm{E}}$, due to the last term in Eq. (4.51). This should be compared with the fact that $\mathrm{Im}\omega$ is finite only for $\omega > \omega_{\mathrm{E}}$ at $T = 0$.

As was shown in Chapter 3, the ratio $2\Delta_0/k_{\mathrm{B}}T_{\mathrm{c}} = 3.53$ and the specific heat jump $\Delta C/C_{\mathrm{n}} = 1.43$ are 'universal' numbers and the ratio $H_{\mathrm{c}}(T)/H_{\mathrm{c}}(0)$ is a 'universal' function (see Eq. (3.28)) in the BCS theory. The strong-coupling corrections to these equations are

$$\frac{2\Delta_0}{k_{\mathrm{B}}T_{\mathrm{c}}} = 3.53\left[1+12.5\left(\frac{k_{\mathrm{B}}T_{\mathrm{c}}}{\omega_{\mathrm{ln}}}\right)^2 \ln\left(\frac{\omega_{\mathrm{ln}}}{2k_{\mathrm{B}}T_{\mathrm{c}}}\right)\right] \quad (4.52)$$

and

$$\frac{\Delta C}{C_{\mathrm{n}}} = 1.43\left[1+53\left(\frac{k_{\mathrm{B}}T_{\mathrm{c}}}{\omega_{\mathrm{ln}}}\right)^2 \ln\left(\frac{\omega_{\mathrm{ln}}}{3k_{\mathrm{B}}T_{\mathrm{c}}}\right)\right] \quad (4.53)$$

according to approximate calculations. These ratios and the comparison of the theoretical $H_{\mathrm{c}}(T)/H_{\mathrm{c}}(0)$ curves with experiments are given in Table 3.1 and Fig. 4.18.

The strong-coupling effect also appears in transport phenomena. The stronger the coupling is, the shorter the quasi-particle lifetime becomes. Consequently the density of states $\mathscr{D}(\omega)$ is modified. We note in particular that because of the appearance of $\mathrm{Im}\Delta$ due to thermal excitations, as was remarked below Eq. (4.51), the divergence of $\mathscr{D}(\omega)$ at the gap edge is weakened and $\mathscr{D}(\omega)$ is broadened as a result. This effect causes suppression of the peak of the nuclear spin relaxation time T_1^{-1} at T_{c}. The degree to which the peak is suppressed depends, of course, on the parameters of the individual system.

Although we have dealt with the cases $\lambda \leq 1$ in the present section, there

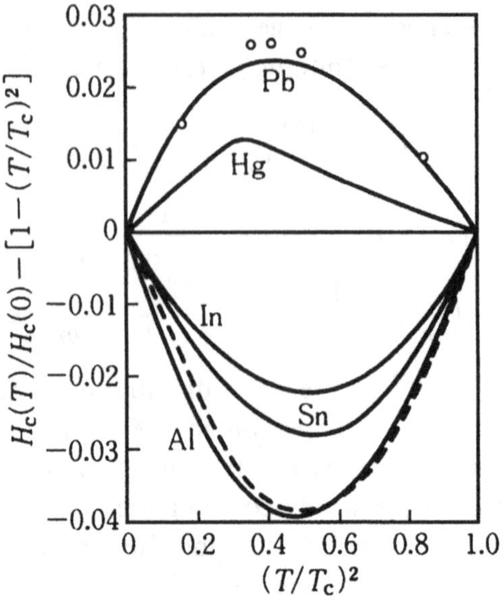

Fig. 4.18. The temperature dependence of the critical field. The broken line is the BCS result while ○ is the strong coupling result in Pb. ([E-2] vol. I.)

are theories of strong-coupling limit with larger λ. We note in particular that the largest possible T_c in the phonon-mediated model has been discussed in connection with high-T_c superconductivity. Similar analysis has been carried out for mechanisms in which bosonic excitations, other than phonons are exchanged. These aspects will be discussed in Chapter 7.

4.6 Impurity effects

The formalism developed here is useful in treating quantitatively the impurity problems considered in Section 3.5, although it is somewhat beyond the scope of the present chapter. It is assumed that the interaction between randomly distributed impurity atoms with positions $\boldsymbol{R}_j$ and the conduction electrons is given by

$$\mathscr{H}_{\text{im}} = \sum \{ V_{\text{N}} a^\dagger_{k'\alpha} a_{k\alpha} + V_{\text{P}} a^\dagger_{k'\alpha} \boldsymbol{\sigma}_{\alpha\beta} a_{k\beta} \cdot \boldsymbol{S}_j \} \mathrm{e}^{\mathrm{i}(\boldsymbol{k}-\boldsymbol{k}')\cdot \boldsymbol{R}_j}. \tag{4.54}$$

Here V_{N} denotes the interaction strength of the spin-independent scattering and V_{P} the strength of the exchange interaction between the impurity atoms with spin $\boldsymbol{S}$ and the conduction electrons. They are both assumed to be of the δ-function (contact) type. The latter interaction represents the paramagnetic

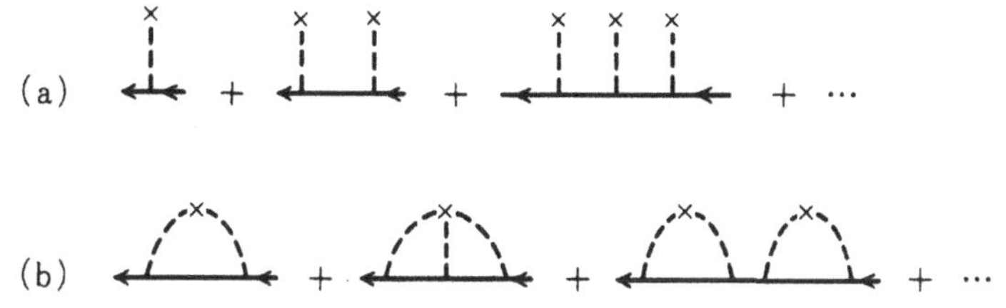

Fig. 4.19. The self-energy diagrams due to impurity scattering.

effect of the impurity atoms. It is assumed that the scattering is elastic and the impurity spin $\boldsymbol{S}$ is classical, that is, it changes only when it interacts with the conduction electrons.

The conduction electron self-energy due to the interaction $\mathscr{H}_{\text{im}}$ is given by the sum of the series shown in Fig. 4.19. After being averaged over the impurity positions, however, only those diagrams given by Fig. 4.19(b) contribute since they are independent of the random phase proportional to $\boldsymbol{R}_j$.

Scattering by a single atom is represented by broken lines starting from a single $\times$. The first diagram of Fig. 4.19(b) is the lowest such process and keeping only this term amounts to the so-called Born approximation. In this approximation, one just replaces the phonon Green's function in our previous analysis by $\cdots \times \cdots$. The dotted line represents V_{N} or V_{P}. Since the scattering is elastic, the energy ω remains unchanged while the momentum and the spin change. Accordingly $\bar{g}(x)$ in Eq. (4.15) is replaced by

$$\bar{g}_{\text{im}}(x) = \frac{1}{2\pi\tau_+}\delta(x), \tag{4.55}$$

where

$$\begin{aligned} \tau_+{}^{-1} &= \tau_{\text{N}}{}^{-1} + \tau_{\text{P}}{}^{-1} \\ \tau_{\text{N}}{}^{-1} &= 2\pi n_{i\text{N}} N(0)\,\overline{|V_{\text{N}}|^2} \\ \tau_{\text{P}}{}^{-1} &= 2\pi n_{i\text{P}} N(0) S(S+1)\,\overline{|V_{\text{P}}|^2}. \end{aligned} \tag{4.56}$$

The average above is taken over the scattering angle near the Fermi surface while $n_{i\text{N}}$ and $n_{i\text{P}}$ denote the concentration of the impurity atoms. From Eqs. (4.55) and (4.14) one obtains in the normal state

$$\Sigma_{\text{im}} = -\mathrm{i}\omega/(2\tau_+|\omega|), \tag{4.57}$$

Fig. 4.20. The self-energy diagrams due to impurity scattering in the superconducting state.

from which one finds that τ_+ is the lifetime of quasi-particles due to impurity scattering.

The Abrikosov–Gor'kov theory (the AG theory) analyses the impurity effects in the superconducting state, using diagrams such as Fig. 4.20 to obtain Σ and Σ_1. Note that the contribution from V_P changes sign in Σ_1 since the spins at each end have opposite sign. Then $\lambda^{(\pm)}$ in Eqs. (4.33) and (4.34) are replaced, when $\omega > 0$, by

$$\begin{aligned} \lambda_{\text{im}}{}^{(+)} &= -\frac{\mathrm{i}}{2\tau_+}\delta(z-\omega) \\ \lambda_{\text{im}}{}^{(-)} &= \frac{\mathrm{i}}{2\tau_-}\delta(z-\omega), \end{aligned} \tag{4.58}$$

where $\tau_-^{-1} = \tau_N^{-1} - \tau_P^{-1}$. In the following, electron–phonon interaction will be treated within the weak-coupling approximation in order to focus on the impurity effects. Accordingly the phonon contribution to Σ is ignored. If Eq. (4.58) is substituted into Eqs. (4.33) and (4.34), one obtains

$$\begin{aligned} [1 - Z(\omega)]\omega &= -\frac{\mathrm{i}}{2\tau_+}\frac{\omega}{\sqrt{\omega^2 - \Delta^2(\omega)}} \\ Z(\omega)\Delta(\omega) &= \bar{\Delta} + \frac{\mathrm{i}}{2\tau_-}\frac{\Delta(\omega)}{\sqrt{\omega^2 - \Delta^2(\omega)}}. \end{aligned} \tag{4.59}$$

Here the weak-coupling approximation was employed to write

$$\bar{\Delta} \equiv \lambda \int_{\Delta_0}^{\omega_c} \mathrm{d}z \; \mathrm{Re}\frac{\Delta(z)}{\sqrt{z^2 - \Delta^2(z)}}. \tag{4.60}$$

From the two equations of Eq. (4.59), we obtain for $u(\omega) \equiv \omega/\Delta(\omega)$

$$\frac{\omega}{\bar{\Delta}} = u\left(1 - \frac{\zeta}{\sqrt{1-u^2}}\right) \tag{4.61}$$

where we have defined

$$\zeta \equiv (\tau_P \bar{\Delta})^{-1}. \tag{4.62}$$

Suppose these impurities were not paramagnetic and $\tau_P = \infty$, that is, $\zeta = 0$. Then from Eq. (4.61) one obtains $\bar{\Delta} = \Delta =$ const. and Eq. (4.60) becomes the same expression as that without impurities. Thus the statement made in Section 3.5 has been justified.

As paramagnetic impurities are increased, not only does the critical temperature T_c decrease by depairing but the *gapless state* appears, where the excitation has no gap in the spectrum even in the superconducting state. One simply needs to inspect Eq. (4.45) to discover this state. Since it follows from Eq. (4.61) that

$$\mathcal{D}(\omega) = N(0)\zeta^{-1}\mathrm{Im}u \tag{4.63}$$

one simply needs to find how $\mathrm{Im}u$ appears as a function of ω in Eq. (4.61). Then one finds that u becomes real for ω less than

$$\omega_g = \bar{\Delta}(1 - \zeta^{2/3})^{3/2}$$

provided that $\zeta < 1$. When $\zeta > 1$, $\mathrm{Im}u$ appears from $\omega = 0$ and hence it is gapless.

It is convenient to use temperature Green's functions to calculate such quantities as $\bar{\Delta}$ and T_c as one may expect from the fact that for an imaginary frequency Eq. (4.61) becomes a relation between real numbers. Here we simply quote the results (see K. Maki in [C-2] vol. II for details).

The change of T_c due to τ_P is given by

$$\ln \frac{T_c}{T_{c0}} = \psi\left(\frac{1}{2}\right) - \psi\left(\frac{1}{2} + \frac{\tau_{Pc}}{\tau_P}\frac{T_{c0}}{T_c}\right), \tag{4.64}$$

where T_{c0} is the critical temperature for $\tau_P = \infty$, ψ is the di-Gamma function and

$$\tau_{Pc}{}^{-1} \equiv \pi T_{c0}/2\gamma = \Delta_{00}/2$$

is the value of τ_P^{-1} at which T_c vanishes. At this value of τ_P^{-1}, the mean free path $\tau_{Pc}v_F$ for spin-flipping scattering becomes of the order of the coherence length ξ_0. Figure 4.21, depicting Eq. (4.64), shows an excellent agreement between experiment and theory.

The supercurrent is stable even in the gapless state and the Meissner effect is also observed there. These facts show that what is essential for superconductivity is not the energy gap but the existence of coherence. We also note firstly, that even non-magnetic scattering causes depairing in the presence of a supercurrent such as in a magnetic field, and secondly,

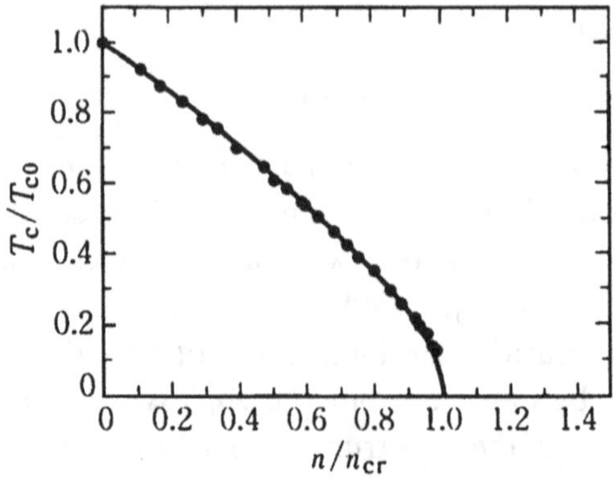

Fig. 4.21. The critical temperature T_c of $(LaGd)Al_2$ as the concentration of the magnetic atom Gd is increased. The temperature T_{c0} defined in the text is 3.24 K while the critical concentration is $n_{cr} = 0.59$ at.%. The solid line shows the result obtained from AG theory.

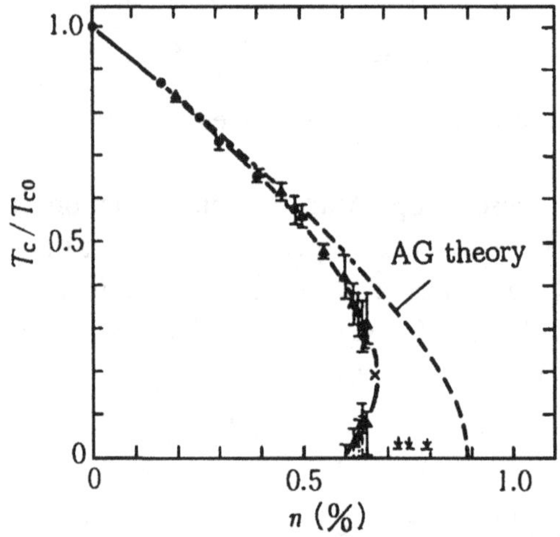

Fig. 4.22. The critical temperature T_c of $(LaCe)Al_2$ as a function of the Ce concentration n. The broken line is obtained from AG theory. (M.B. Maple in [D-5].)

that pairs are broken even by non-magnetic scattering except in an s-wave superconductor.

Although the spin of the impurity atom has been treated classically in the analyses so far, the Kondo effect appears for $V_P > 0$ (antiferromagnetic coupling) if the process given by the second diagram in Fig. 4.19(b) is considered. The effect of an impurity atom which shows the Kondo effect manifests itself clearly when the Kondo temperature T_K is close to T_c. In this case, there is a conflict between a bound state between conduction electrons (i.e., the Cooper pair) and a singlet bound state between a conduction electron and an impurity spin. Therefore T_c becomes zero at a relatively low impurity density. Moreover, since τ_P^{-1} changes with temperature when $T_c > T_K$, there appears a re-entrant phenomenon, shown in Fig. 4.22, in which the sample becomes a superconductor at low temperature but it becomes a normal conductor again as temperature is lowered further.

5

Ginzburg–Landau theory

The Ginzburg–Landau (GL) theory has been employed for general analysis of the physical properties of an ordered phase. This theory is especially useful when the system under consideration has a spatial variation, due to an applied external field for example. It should also be noted that this theory serves as a basis for studying the effects of fluctuations, that are neglected in the mean-field approximation. The GL theory was proposed before the microscopic theory of superconductivity by BCS and has played an important rôle in the analysis of superconducting phenomena even after the emergence of BCS theory. In the present chapter, the time-independent GL theory and its applications are first discussed; its extensions and the effect of fluctuations are then considered.

5.1 GL theory of superconductivity

The free energy F_s of the system is not only a function of the temperature T and other ordinary thermodynamic variables but a function (in fact, a *functional* in general since we are dealing with a system with spatial variations) of the order parameter. The order parameter of a spin-singlet s-wave superconductor is the pair wave function $\Psi(\boldsymbol{x}) = \langle \psi_\uparrow(\boldsymbol{x})\psi_\downarrow(\boldsymbol{x})\rangle$, which is equivalent to the gap parameter $\Delta(\boldsymbol{x}) = g\Psi(\boldsymbol{x})$ in the BCS model. Here the order parameter is written $\Psi(\boldsymbol{x})$ including a multiplicative constant. Let us write the free energy as $F_s = F_s(\{\Psi(\boldsymbol{x})\})$, where such variables as T and V are and will be dropped to simplify the notation. It is assumed, in the GL theory, that the spatial variation is gradual and, accordingly, the free energy density f_s is taken to be a function of $\Psi(\boldsymbol{x})$ and $\nabla\Psi(\boldsymbol{x})$ only. Moreover, since the free energy is left invariant under gauge transformations, which change the phase of Ψ, and spatial rotations, f_s must be a function of the combinations $\Psi^*(\boldsymbol{x})\Psi(\boldsymbol{x})$ and $\nabla\Psi^*(\boldsymbol{x}) \cdot \nabla\Psi(\boldsymbol{x})$ only. Actual calculations are possible only when $|\Psi|^2$ and $|\nabla\Psi|^2$ are small and hence f_s can be expanded in powers of these quantities. As a result, a commonly used form of the free energy in the GL theory is

$$F_s = \int dx \left\{ -a|\Psi|^2 + \frac{b}{2}|\Psi|^4 + c|\nabla\Psi|^2 \right\} + F_n, \tag{5.1}$$

which is applicable when T is close to T_c. F_n is identified with the free energy of the normal state since the order parameter Ψ vanishes for this state.

First note that the order parameter $\Psi = \Psi_e$ is realised in a uniform equilibrium state ($\nabla\Psi = 0$). Thus one has $|\Psi_e|^2 = a/b$, for which $(F_s - F_n)_e = -a^2/2b = -H_c^2(T)/8\pi$, see Section 3.2. If one rewrites $\Psi/|\Psi_e|$ as Ψ, the free energy relative to the normal state is written as

$$F_s - F_n = \frac{H_c^2(T)}{4\pi} \int dx \left\{ -|\Psi|^2 + \frac{1}{2}|\Psi|^4 + \xi^2|\nabla\Psi|^2 \right\}, \tag{5.2}$$

where the coherence length has been defined by $\xi(T) \equiv \sqrt{c/a}$.

The coefficients of the above equation in the BCS model have been obtained in Section 3.2. Since the gradient energy with the order parameter $\Psi \propto e^{i\boldsymbol{q}\cdot\boldsymbol{x}}$ must be the same as Eq. (3.41), one finds

$$\xi(T) = v_F/\sqrt{6}\Delta_e(T) = 0.74\xi_0(1 - T/T_c)^{-1/2}, \tag{5.3}$$

where ξ_0 is defined by Eq. (3.16) at $T = 0$. (It should be noted that $\xi(T)$ here differs from the definition used in Eq. (3.16) by a factor of $\pi^2/6$.) When the mean free path l is finite due to the presence of non-magnetic scattering, ξ^2 above should be multiplied by a factor

$$\eta \equiv \frac{8}{7}\zeta(3) \sum_{n=0}^{\infty} \left[(2n+1)^2(2n+1+\xi_0/l) \right]^{-1} \sim (1 + 0.752\xi_0/l)^{-1}.$$

Let us next consider a system coupled with a static magnetic field. Since Ψ is the wave function of a charged pair and the free energy should be invariant under gauge transformations of the second kind (local gauge transformations), the third term of Eq. (5.2) should take the form

$$\xi^2 \left(-\frac{\nabla}{i} - \frac{e^*}{\hbar c}\boldsymbol{A} \right) \Psi^* \cdot \left(\frac{\nabla}{i} - \frac{e^*}{\hbar c}\boldsymbol{A} \right) \Psi.$$

Although e^* is arbitrary in a phenomenology, we will take the pair charge $e^* = 2e$ as in the BCS model. Supplementing the magnetic field energy

$\int d\boldsymbol{x}\boldsymbol{B}^2/8\pi$, one finally obtains the free energy

$$F = \frac{H_c^2(T)}{4\pi}\int d\boldsymbol{x}\left\{-|\Psi|^2 + \frac{1}{2}|\Psi|^4 + \xi^2\left(-\frac{\nabla}{\mathrm{i}} - \frac{2\pi}{\phi_0}\boldsymbol{A}\right)\Psi^* \cdot \left(\frac{\nabla}{\mathrm{i}} - \frac{2\pi}{\phi_0}\boldsymbol{A}\right)\Psi\right\} + \frac{1}{8\pi}\int d\boldsymbol{x}(\nabla\times\boldsymbol{A})\cdot(\nabla\times\boldsymbol{A}), \tag{5.4}$$

where $\phi_0 = hc/2e$ is the flux quantum introduced in Section 1.1.

The field configurations $\Psi(\boldsymbol{x})$ and $\boldsymbol{A}(\boldsymbol{x})$ in an equilibrium state are determined by the conditions $\delta F/\delta\Psi^* = 0$ and $\delta F/\delta \boldsymbol{A} = 0$. Explicit expressions of these conditions are

$$\xi^2\left(\frac{\nabla}{\mathrm{i}} - \frac{2\pi}{\phi_0}\boldsymbol{A}\right)^2\Psi - \Psi + |\Psi|^2\Psi = 0 \tag{5.5}$$

$$\nabla\times(\nabla\times\boldsymbol{A}) = \frac{4\pi}{c}\boldsymbol{j}_s \tag{5.6}$$

$$\boldsymbol{j}_s(x) = \frac{e n_s(T)}{4m}\frac{1}{\mathrm{i}}\{\Psi^*\cdot\nabla\Psi - \nabla\Psi^*\cdot\Psi\} - \frac{e^2}{mc}n_s(T)|\Psi|^2\boldsymbol{A}, \tag{5.7}$$

where $n_s(T)$ is the density of the superconducting component defined by Eq. (3.40), which reduces to $n_s(T) = 2n(1 - T/T_c)$ in the vicinity of T_c. If the number density of the pairs $n_s/2$ is employed, one finds that the current density (5.7) is a quantum mechanical expression for a particle of mass $2m$ and charge $2e$. Equation (5.5) is called the *Ginzburg–Landau equation.* The analogy between this equation and Eq. (1.7) should be obvious.

In addition to the coherence length ξ in Eq. (5.5), there appears another characteristic length scale λ in Eq. (5.6), defined by

$$\lambda^{-2} \equiv 4\pi e^2 n_s(T)/mc^2,$$

which is just the London penetration depth given by Eq. (1.12). Equation (5.6) is then rewritten as

$$\lambda^2\nabla\times(\nabla\times\boldsymbol{A}) = (\phi_0/4\pi\mathrm{i})(\Psi^*\cdot\nabla\Psi - \nabla\Psi^*\cdot\Psi) - |\Psi|^2\boldsymbol{A}. \tag{5.8}$$

The ratio of the two characteristic lengths

$$\kappa \equiv \lambda/\xi = \sqrt{2}\,2\pi H_c(T)\lambda^2/\phi_0 \tag{5.9}$$

is called the *GL parameter*. We note *en passant* that λ^2 has a factor η^{-1} when l is finite, and hence $\kappa \propto \eta^{-1}$.

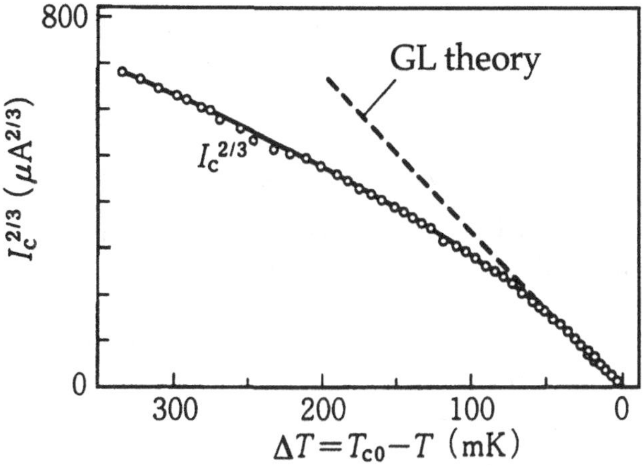

Fig. 5.1. The temperature dependence of the critical current I_c measured on a thin wire of Sn.

5.1.1 Transition in the presence of a supercurrent

Let us consider, as the simplest application of the GL theory, the transition in a thin wire carrying a supercurrent. If the radius of the wire is less than both $\xi(T)$ and $\lambda(T)$ introduced above, then Ψ is considered to be constant across the cross-section and the effect of a magnetic field is negligible. Therefore the problem reduces to a one-dimensional problem, whose coordinate is written as x. If one writes $\Psi = \psi e^{ikx}$, ψ being a real constant, one obtains $\psi^2 = 1 - \xi^2 k^2$ from Eq. (5.5) and accordingly, from Eq. (5.7), $j_x = (n_s/2m)k(1 - \xi^2 k^2)$. This supercurrent takes the maximum value

$$j_c = \frac{en_s}{3\sqrt{3}m\xi} = \frac{(c^2/4\pi e)}{3\sqrt{3}\lambda^2\xi} \tag{5.10}$$

when $\xi k = 1/\sqrt{3}$. If $F \propto (1 - \xi^2 k^2)^2$ is looked upon as a function of j, it is found that $\partial F/\partial j$ diverges at j_c. Thus it follows that j_c is the critical current in the sense that it is the maximum possible current without destroying superconductivity. The system makes a first order transition to a normal state if a current $j > j_c$ flows.

It follows from Eq. (5.10) that $j_c \propto (1 - T/T_c)^{3/2}$. Figure 5.1 is the $I - V$ characteristic of a whisker of Sn showing the current at which the voltage across the sample appears, as a function of temperature. A close study of the data reveals that there is a deviation from the theory near T_c, which is considered to be due to a thermal fluctuation effect. On the other hand, the system cannot be regarded as one-dimensional at low temperatures and there is a deviation from the $I^{3/2}$-rule. In fact, the $I - V$ characteristic of a thin wire is a complicated subject and is an interesting problem which cannot be explained by the simple transition above, see Section 5.6.

5.1.2 Transition in a cylinder of a thin film in a magnetic field

Let us consider the transition in a cylinder of a thin film placed in a uniform magnetic field parallel to the axis of the cylinder. It is assumed that the radius R of the cylinder is much larger than λ and the film thickness is smaller than λ. Accordingly the magnetic field is constant over the whole sample and, since $B = H$, the vector potential $\boldsymbol{A}$ has only the ϕ-component in cylindrical coordinates, which is taken as $A_\phi = RH/2$ in the film. If $\Psi = \psi e^{in\phi}$ is substituted into Eq. (5.5), one finds

$$\xi^2 \left(\frac{n}{R} - \frac{\pi RH}{\phi_0} \right)^2 \psi - (1 - \psi^2)\psi = 0.$$

(Note that one is lead to the magnetic flux quantisation for a thick film since the first term must then vanish.) In the present case the integer n, which minimises the free energy, is realised of course. The deviation of the critical temperature $\Delta T_c = T_c - T_{c0}$, where T_{c0} is the critical temperature at $H = 0$, is therefore given by

$$\Delta T_c / T_c = -0.548(\xi_0/R)^2 (n - \Phi/\phi_0)^2, \tag{5.11}$$

which is a periodic function of the flux Φ through the cylinder. In contrast to j_c of the previous problem, there is a second order transition at T_c since the external field is related only to the phase of Ψ. Figure 5.2 shows the relation between T_c and H determined by the temperature dependence of the resistance, and clearly indicates the periodic dependence of T_c on the flux. The thickness of the film must be considered to explain why the maxima in T_c become smaller as H increases.

5.2 Boundary energy

Let us consider the system shown in Fig. 5.3, where the plane $x = 0$ is the boundary between a superconducting state ($x \to +\infty$) and a normal state with an applied critical magnetic field ($x \to -\infty$). One may choose the gauge $\boldsymbol{A} = (0, A(x), 0)$ provided that $\boldsymbol{B} \parallel \hat{z}$. While we can find Ψ and B by solving Eqs. (5.5) and (5.6), we are interested in the free energy as a function of an external magnetic field, given by $G = F - \int dx \boldsymbol{H} \cdot \boldsymbol{B}/4\pi$. Suppose $\Psi = 1, B = 0$ for $x > 0$ and $\Psi = 0, B = H_c$ for $x < 0$. Then the

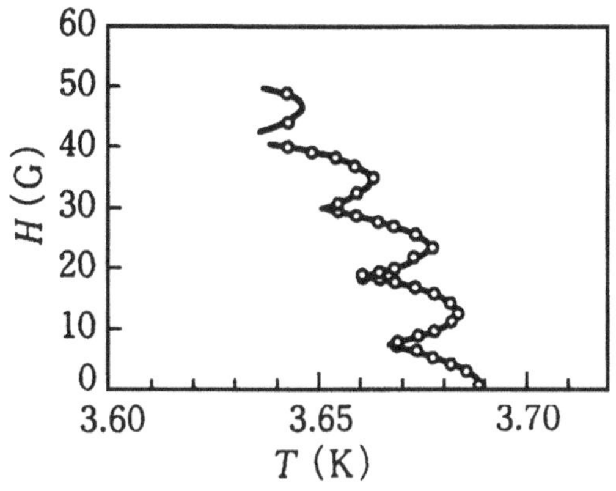

Fig. 5.2. The critical temperature as a function of the flux through a cylinder of a thin film of Sn. (From [G-13].)

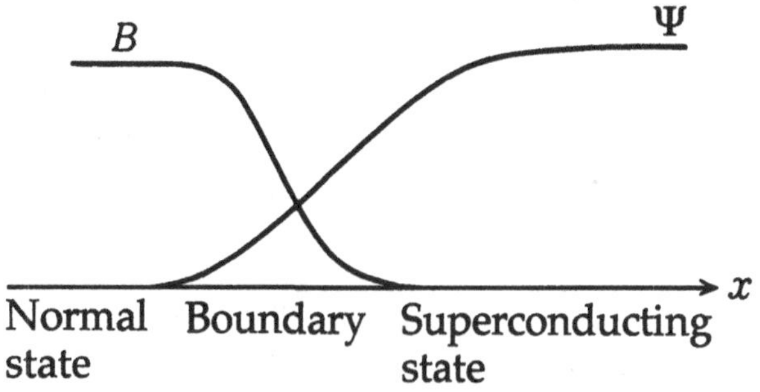

Fig. 5.3. Variations of the magnetic field and the order parameter at the boundary of a superconducting state and a normal state.

density of the free energy G is $-H_c^2/8\pi$ everywhere. Therefore the difference between the free energy G of the configuration shown in Fig. 5.3 and this free energy is

$$\begin{aligned} E_b &= F_{GL} + \int_{-\infty}^{\infty} dx \left(\frac{B^2}{8\pi} - \frac{BH_c}{4\pi} \right) - \int_{-\infty}^{\infty} dx \left(-\frac{H_c^2}{8\pi} \right) \\ &= \frac{H_c^2}{4\pi} \int_{-\infty}^{\infty} dx \left\{ \xi^2 \left| \left(\frac{\nabla}{i} - \frac{2\pi}{\phi_0} A \right) \Psi \right|^2 \right. \\ &\quad \left. - \left(1 - \frac{1}{2}|\Psi|^2 \right) |\Psi|^2 + \frac{1}{2} \left(\frac{B}{H_c} - 1 \right)^2 \right\}. \end{aligned} \tag{5.12}$$

Furthermore, if one subtracts $\int_{-\infty}^{\infty} dx \Psi^* \{\text{GL equation}\} = 0$ from the above

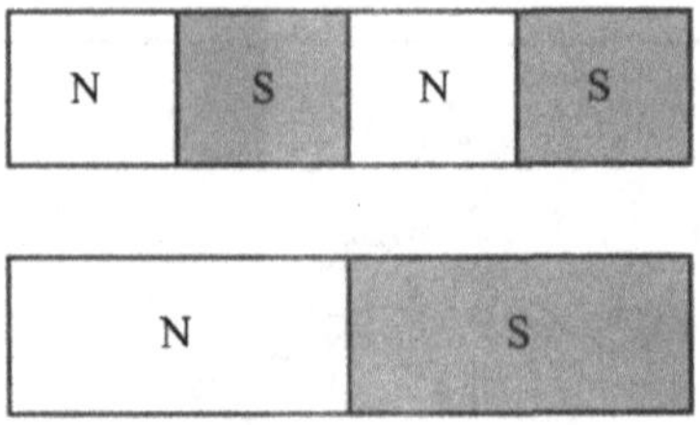

Fig. 5.4. The boundary energy $2E_b$ is the free energy difference between the two states.

equation, one obtains a simple expression

$$\frac{E_b}{H_c^2/8\pi} = \int_{-\infty}^{\infty} dx \left\{ \left(1 - \frac{B}{H_c}\right)^2 - |\Psi|^4 \right\}, \tag{5.13}$$

since Ψ is a solution to Eq. (5.5). It is reasonable to consider the boundary energy thus defined because the free energy difference between the two states shown in Fig. 5.4 is written in terms of E_b as $3E_b - E_b = 2E_b$.

One has to solve Eqs. (5.5) and (5.6) to find E_b, which is easy if $\Psi = \Psi(x)$ is chosen to be real. Let us scale lengths by λ and vector potentials by $\phi_0/2\pi\lambda$, that is:

$$x \to x/\lambda, \quad A \to A/(\phi_0/2\pi\lambda). \tag{5.14}$$

Then the equations to be solved are

$$\begin{gathered} \kappa^{-2}\left[\frac{d^2}{dx^2} - A^2(x)\right]\Psi + (1-\Psi^2)\Psi = 0 \\ \frac{d^2 A}{dx^2} = A\Psi^2. \end{gathered} \tag{5.15}$$

The boundary conditions are $\Psi = 1, A = 0$ as $x \to \infty$ and $\Psi = 0, dA/dx = \kappa/\sqrt{2}$ as $x \to -\infty$, where use has been made of Eq. (5.9).

Let us first consider the limit $\kappa \ll 1$. Since A is of the order of κ, the first equation of (5.15) is approximated by $-(1/\kappa^2)d^2\Psi/dx^2 - (1-\Psi^2)\Psi = 0$ at $x \gg 1$ and, as a consequence, the solution is given by $\Psi = \tanh(\kappa x/\sqrt{2})$. The magnetic field penetrates into the sample to a distance λ. Therefore one obtains

$$\frac{E_b}{H_c^2/8\pi} \simeq \lambda \int_0^{\infty} dx(1-\Psi^4) = \frac{4\sqrt{2}}{3}\frac{\lambda}{\kappa}. \tag{5.16}$$

Let us next consider the opposite limit $\kappa \gg 1$, namely the London limit. The magnetic field penetrates toward the interior of the sample in the form e^{-x} and Ψ is restored as $1 - e^{-\kappa x}$. Therefore Eq. (5.13) is estimated to be $\sim -3\lambda/2$. In this way, boundary energy changes sign according to the value of κ. In fact, it can be proved from Eq. (5.15) that E_b vanishes when $\kappa = 1/\sqrt{2}$, see [A-2].

In summary, superconductors are classified according to whether $\kappa < 1/\sqrt{2}$ (type 1 superconductors) or $\kappa > 1/\sqrt{2}$ (type 2 superconductors). The boundary energy E_b is negative when $\kappa > 1/\sqrt{2}$, in which case it is expected that the free energy is lowered if there are as many S–N boundaries as possible, as shown in Fig. 5.4. Accordingly the superconducting states in a magnetic field differ significantly depending on whether they are of type 1 or type 2. There are two critical magnetic fields $H_{c1} < H_{c2}$ in a type 2 superconductor, as is shown in the next section, and a vortex lattice state is realised when a magnetic field intermediate between the two critical fields is applied.

5.3 Critical fields H_{c1} and H_{c2}

The magnetisation curve of a type 2 superconductor, for which the GL parameter κ is greater than $1/\sqrt{2}$, has been shown in Fig. 1.4(b). A magnetic flux starts to penetrate into a sample as an external magnetic field H exceeds the lower critical field $H_{c1}(T)$, which is less than the thermodynamic critical magnetic field $H_c(T)$. The sample makes a transition to a normal state as the field H is further increased beyond the upper critical field $H_{c2}(T)$, which is greater than $H_c(T)$.

5.3.1 *Lower critical field* H_{c1}

When a magnetic field is applied parallel to the axis of a long superconducting cylinder, the first flux penetrating the sample should be located along the axis of the cylinder. Therefore, the vector potential in cylindrical coordinates takes the form $A = (0, A_\phi(r), 0)$ and consequently the magnetic field is given by $B_z(r) = [\partial(rA_\phi)/\partial r]/r$. (The subscript indicating the component will be omitted in the following.) Since the supercurrent is an axially symmetric vortex current, the order parameter may be put in the form $\Psi = \psi(r)\mathrm{e}^{in\phi}$, where ψ is real. Substituting these expressions into Eqs. (5.5) and (5.6) one obtains

$$-\frac{1}{\kappa^2}\left\{\frac{1}{r}\frac{\mathrm{d}}{\mathrm{d}r}\left(r\frac{\mathrm{d}\psi}{\mathrm{d}r}\right) - \left(\frac{n}{r} - A\right)^2\psi\right\} - (1-\psi^2)\psi = 0 \tag{5.17}$$

$$-\frac{\mathrm{d}}{\mathrm{d}r}\frac{1}{r}\frac{\mathrm{d}}{\mathrm{d}r}(rA) = \left(\frac{n}{r} - A\right)\psi^2, \tag{5.18}$$

where A has been scaled by $2\pi H_c\lambda^2/\phi_0$ to be made dimensionless.

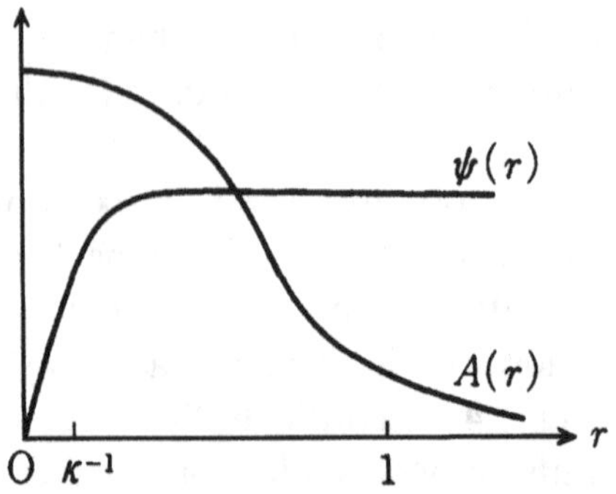

Fig. 5.5. The spatial variations of the order parameter ψ and the vector potential A associated with a quantised flux.

The above equations can be treated approximately when $\kappa \gg 1$, in which case the scale of variation of $A(r)$ is of the order of unity while that of $\psi(r)$ is $\sim 1/\kappa$ (that is, of the order of λ and ξ respectively in physical units) as is shown in Fig. 5.5. Therefore one may put $\psi(r) \simeq 1$ in the region $r > \kappa^{-1}$. Then Eq. (5.18) reduces to

$$\left\{\frac{\mathrm{d}^2}{\mathrm{d}r^2} + \frac{1}{r}\frac{\mathrm{d}}{\mathrm{d}r} - \left(1 + \frac{1}{r^2}\right)\right\}\left(\frac{n}{r} - A\right) = 0.$$

Since A should not blow up at $1 \gg r$ ($\gg 1/\kappa$), the solution must be proportional to $1/r$. Such a solution assumes the form

$$\frac{n}{r} - A = nK_1(r), \tag{5.19}$$

where $K_1(z) = -(\pi/2)[J_1(\mathrm{i}z) + \mathrm{i}N_1(\mathrm{i}z)]$ is the modified Bessel function. (Note that $K_1(z) \to 1/z$ as $|z| \to 0$.) When $r \gg 1$, on the other hand, $K_1(r)$ behaves like $\mathrm{e}^{-r}/\sqrt{r}$ and accordingly the magnetic flux is confined within the region $r \lesssim 1$. Therefore the total flux through a circle of radius $R \gg 1$ is quantised as

$$\int_{r \le R} \boldsymbol{B}(r) \cdot \mathrm{d}\boldsymbol{S} = \int_{r=R} \boldsymbol{A} \cdot \mathrm{d}\boldsymbol{l} = 2\pi n \ (= n\phi_0). \tag{5.20}$$

Let us next consider the difference between the free energy G of the above configuration and G_0 of the configuration $\psi = 1, B = 0$ to find the lower critical field $H_{\mathrm{c}1}$. If the method used to obtain Eq. (5.13) is employed again, the free energy difference per unit length is expressed as

$$\begin{aligned}\Delta g &\equiv (G - G_0)/(H_{\mathrm{c}}^2(T)/4\pi) \\ &= \frac{1}{2}\int \mathrm{d}x(1 - |\psi|^4) + \frac{2}{\kappa^2}\int \mathrm{d}x\left\{\frac{1}{2}(H - B)^2 - \frac{1}{2}H^2\right\}. \end{aligned} \tag{5.21}$$

In the region $r > 1/\kappa$, where $A(r)$ is given by Eq. (5.19), the derivative term in Eq. (5.17) is negligible and, combining it with the Maxwell equation (5.18), one may put $(1 - \psi^2) \simeq \kappa^{-2}(\nabla \times \boldsymbol{B})^2$. If this relation is substituted into Eq. (5.21), one obtains, along with the estimate $|\psi|^2 = 1 + O(1/\kappa^2)$,

$$\Delta g = \frac{1}{\kappa^2} \int \mathrm{d}x \{(\nabla \times \boldsymbol{B})^2 + B^2 - 2HB\}. \tag{5.22}$$

Since one obtains from Eq. (5.18) $\nabla \times (\nabla \times \boldsymbol{B}) \sim -\boldsymbol{B}$ in the same region, the sum of the first and the second terms in the curly brackets above may be written in the divergence form as $\nabla(\boldsymbol{B} \times (\nabla \times \boldsymbol{B}))$. Then the integral is estimated as

$$-\frac{1}{\kappa^2} \int_{r=\kappa^{-1}} r\mathrm{d}\phi \; (\boldsymbol{B} \times (\nabla \times \boldsymbol{B}))_r = 2\pi\kappa^{-2}(\ln \kappa + 0.116),$$

where use has been made of the relation $B(r) = nK_0(r)$ and the asymptotic form $K_0(r) \simeq -\ln r + \ln 2 - \gamma$. The third term is found to be equal to $-4\pi nH/\kappa^2$ from Eq. (5.20). Therefore one finds

$$\Delta g \simeq 2\pi\kappa^{-2}(\ln \kappa + 0.116) - 4\pi n\kappa^{-2}H. \tag{5.23}$$

The smallest H which makes the above expression negative corresponds to the lower critical field. If Eq. (5.9) is used to restore the dimensions, one obtains

$$H_{c1}(T)/H_c(T) = (1/\sqrt{2}\kappa)[\ln \kappa + 0.116]. \tag{5.24}$$

That is, a vortex with a flux quantum ϕ_0 penetrates into the sample at $H = H_{c1}$. Note that Eq. (5.24) is applicable only when $\kappa \gg 1$ is satisfied. Note also that $\psi(r) \propto r$ in the vicinity of $r = 0$ when $n = 1$. The central part of the vortex with $0 \leq r \lesssim \xi$, where the order parameter changes from 0 $(r = 0)$ to ~ 1 $(r \simeq \xi)$, is called the vortex core. The contribution to the energy from the core is of the order of κ^{-2}.

The free energy of two vortices placed at a finite distance $r \gg \kappa^{-1}\lambda$ is larger than that with $r \to \infty$ by $(\phi_0^2/16\pi^2\lambda^2)K_0(r/\lambda)$, which means that a repulsive force acts between two vortices. Due to a screening effect, however, the repulsion weakens exponentially at $r > \lambda$. As a result a magnetic flux suddenly starts to penetrate into a sample as H is increased above H_{c1}.

5.3.2 Upper critical field H_{c2}

A type 2 superconductor under a sufficiently strong uniform magnetic field is in a normal state. As H is decreased, superconductivity appears, that is, the order parameter Ψ becomes finite at a certain value of H, which we call the upper critical field H_{c2}. At this field, therefore, $|\Psi|$ is small throughout the sample and the effect of supercurrent is negligible. In other words, the theory may be linearised with respect to $|\Psi|$ and one may put $B = H$ everywhere. Let us suppose $\boldsymbol{H} \parallel \hat{\mathbf{z}}$ and choose the gauge $A_x = A_z = 0$, $A_y = Hx$. We need to consider Eq. (5.5) only, which may be written as

$$-\frac{1}{\kappa^2}\left[\frac{\partial^2}{\partial x^2} + \left(\frac{\partial}{\partial y} - \mathrm{i}\frac{x}{{r_0}^2}\right)^2\right]\psi = (1+\omega)\psi, \tag{5.25}$$

where

$${r_0}^{-2} \equiv 2\pi H\lambda^2/\phi_0. \tag{5.26}$$

Equation (5.25) is written in the form of an eigenvalue equation for convenience. A solution $\psi \neq 0$ appears when $\omega \leq 0$. This equation is just the Schrödinger equation in the presence of a uniform magnetic field, whose solutions are the Landau orbitals.

If one puts $\psi = \mathrm{e}^{\mathrm{i}py} f(x)$, Eq. (5.25) becomes

$$-\kappa^{-2}(D_{(+)}D_{(-)} - {r_0}^{-2})f = (1+\omega)f \tag{5.27}$$

where

$$D_{(\pm)} \equiv \left(\frac{\mathrm{d}}{\mathrm{d}x'} \mp \frac{x'}{{r_0}^2}\right), \quad x' = x - p{r_0}^2.$$

This is the Schrödinger equation of a harmonic oscillator with the centre pr_0^2 and the solution is given by

$$\begin{aligned} f_0 &\propto \exp(-x'^2/2{r_0}^2), \quad f_n \propto D^n_{(+)} f_0 \\ \omega_n &= (2n+1)(\kappa r_0)^{-2} - 1. \end{aligned} \tag{5.28}$$

Thus ω_0 is found to cross zero when $\kappa r_0 = 1$ and as a result, superconductivity emerges at

$$H = H_{c2} = \phi_0/2\pi\xi^2 = \sqrt{2}\,\kappa H_c(T). \tag{5.29}$$

5.4 Vortex lattice states

The phase diagram of a conventional type 2 superconductor is shown in Fig. 5.6. It had been thought that the line H_{c2} dividing the shaded area and the normal state in this figure corresponded to a lattice of vortex lines carrying a flux quantum. Recently, however, during research on high-T_c superconductors with extremely small coherence length, it was found that the line $H_{c2}(T)$ is not a clear phase boundary and that the vortex lattice just below the $H_{c2}(T)$ line is, rather, a liquid state continuously connected

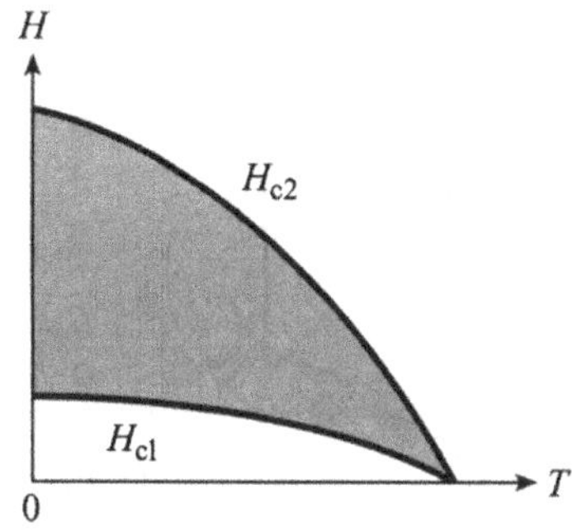

Fig. 5.6. Phase diagram of a type 2 superconductor.

to the normal state. This is in accord with theoretical analysis which takes into account fluctuation of the order parameters (see Chapter 7 and Section 2 in Appendix A2). In spite of this recent progress, it is still necessary to know the results of the mean-field theory (the GL theory), to understand the properties of type 2 superconductors. A.A. Abrikosov was the first to study the lattice state near $H_{c2}(T)$, see [G-14].

Since $H \sim H_{c2}(T)$, one may consider that $|\Psi| \ll 1$ and accordingly the nonlinear term in the GL equation is small and also that the deviation of the flux density B from a uniform B_0 in the superconductor is small. Therefore Ψ may be approximated by a superposition of the lowest energy solutions

$$\psi \propto e^{ipy} f_0(x') = \exp\left[ipy - \frac{1}{2}\left(\frac{x}{r_0} - pr_0\right)^2\right]$$

of the linearised GL equation in the presence of a magnetic field B_0. One should recall here that $r_0^{-2} = 2\pi B_0 \lambda^2/\phi_0$. Keeping in mind that p is the centre of the Gaussian function ψ, let us consider the superposition

$$\phi(x, y) = C \sum_{l=-\infty}^{\infty} \exp\left\{-\frac{1}{2}\left(\frac{x}{r_0} - \frac{2\pi}{a}l\right)^2 + i\frac{2\pi}{a}l\left(\frac{y}{r_0} - \frac{b}{2a}l\right)\right\}. \tag{5.30}$$

This form of ϕ clearly has the following periodic properties:

$$\begin{aligned} &\phi(x, y + ar_0) = \phi(x, y) \\ &\phi(x + 2\pi r_0/a, y + br_0/a) = \phi(x, y) \cdot \exp[i(2\pi/a)(y/r_0 + b/2a)]. \end{aligned} \tag{5.31}$$

(In fact, ϕ can be written in terms of Jacobi's ϑ function and the above periodicity is a direct consequence of this fact.) Therefore ψ is periodic, if we disregard its phase, with the unit cell shown in Fig. 5.7. The area of the unit cell $(2\pi/a)r_0 \times ar_0 = 2\pi r_0^2$ is pierced through by exactly a unit flux quantum

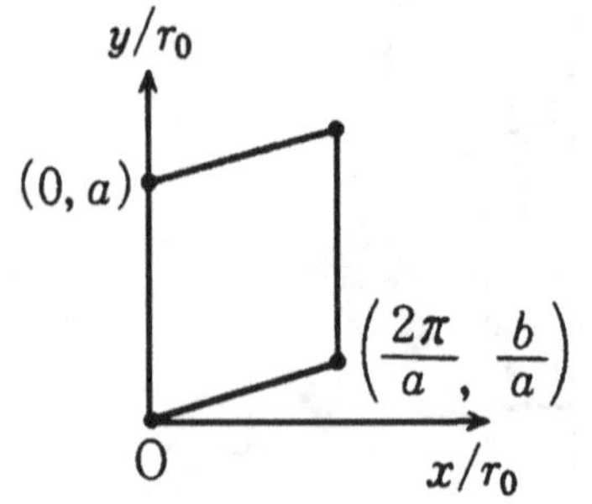

Fig. 5.7. The unit cell of a flux lattice.

Fig. 5.8. The contour lines of $|\phi|^2$ in a triangular lattice. C denotes the centre and M is a maximal point while S is a saddle point.

and $\phi(\pi r_0/a, br_0/a) = 0$. If the choice $b/a = a/2 = 3^{-1/4}\sqrt{\pi}$ is made, the resulting lattice is a triangle whose edge has the length $2\sqrt{\pi/\sqrt{3}}$. The pattern of the vortex current, which is also the contour lines of $|\phi|^2$, is shown in Fig. 5.8.

The free energy of the above system including the magnetic energy generated by the supercurrent is

$$\frac{F}{V} = \left\{\frac{1}{\kappa^2 r_0{}^2} - 1 + \frac{1}{4\kappa^2}\left[(2\kappa^2 - 1)\beta_{\mathrm{A}} + 1\right]\overline{|\phi|^2}\right\} \times \overline{|\phi|^2}$$

where V is the system volume, $\overline{|\phi|^2}$ is the spatial average of $|\phi|^2$ and $\beta_{\mathrm{A}} \equiv \overline{|\phi|^4}/(\overline{|\phi|^2})^2$. $\overline{|\phi|^2}$ is determined so as to minimise the free energy above and as a result one obtains the minimum value

$$\frac{F}{V} = -\frac{1}{8\pi}(B_0 - H_{c2})^2 \frac{1}{(2\kappa^2 - 1)\beta_A + 1}, \tag{5.32}$$

where the physical dimensions are recovered. One finds from the above expression that the lattice giving the smallest β_A is most stable. This is found to be the triangular lattice, for which $\beta_A = 1.16$. Since $-4\pi M = B_0 - H$, the magnetisation is given by

$$M = -(H_{c2} - H)/[4\pi(2\kappa^2 - 1)\beta_A],$$

see [A-2] for details.

5.4.1 *Quasi-particle excitations associated with a vortex*

Figure 1.6 is an STM (scanning tunnelling microscope) image of a vortex lattice obtained by scanning a surface perpendicular to the applied magnetic field. The sample is a layered crystal $NbSe_2$ ($T_c = 7.2$ K, $\kappa \simeq 9$) and a magnetic field of 1 T was applied perpendicular to the layers. Since a tunnelling current between the needle and the sample surface is observed in the STM, this system can be treated in the same way as the tunnelling current between a superconductor and a normal metal studied in Section 3.4. Since a vortex lattice state is not spatially uniform, however, the quasi-particle spectrum changes with the nonuniformity and thus an image such as Fig. 1.6 is obtained. Theoretically a space-dependent quasi-particle density of states

$$\mathscr{D}(\omega) = \sum_n [\,|u_n(\boldsymbol{x})|^2 - |v_n(\boldsymbol{x})|^2]\delta(\omega - \omega_n), \tag{5.33}$$

necessary to find the tunnelling current, may be obtained from the solutions $u_n(x)$ and $v_n^*(x)$ to the Bogoliubov–de Gennes (BG) equation (see Section 3.5).

Let us consider a superconductor with $\kappa \gg 1$, as before, whose impurity effects are negligible. An external field H is taken to be sufficiently smaller than H_{c2} and accordingly it may be assumed that the lattice constant of the vortex lattice a is larger than ξ and that each vortex line is independent of the others. Consider, therefore, a single vortex along the z-axis and introduce the cylindrical coordinates (r, ϕ, z). Although a vector potential appears in the BG equations, its effect is negligible since $A_\phi \sim 2\pi\phi_0 r/a^2$ in $r \lesssim \xi$, which is an important region in the following discussions (note that $B \sim \phi_0/a^2$). Since the order parameter takes the form $\psi(r)e^{i\phi}$, the energy gap parameter also takes the form $\Delta = \Delta(r)e^{i\phi}$ with $\Delta(r)$ real. Therefore one puts

$$u_n(\boldsymbol{x}) = u_{lk}(r)e^{ikz+il\phi}$$
$$v_n^*(\boldsymbol{x}) = v_{lk}^*(r)e^{ikz+i(l-1)\phi},$$

where l is an integer and the radial quantum number is omitted. Substituting the above expressions into the BG equations, one obtains

$$\begin{aligned}
&[-D + l^2/r^2 - k_F{}^2 \sin^2\alpha - 2m\varepsilon]u_{lk}(r) + 2m\Delta(r)v_{lk}{}^*(r) = 0 \\
&[-D + (l-1)^2/r^2 - k_F{}^2 \sin^2\alpha + 2m\varepsilon]v_{lk}{}^*(r) - 2m\Delta(r)u_{lk}(r) = 0,
\end{aligned} \tag{5.34}$$

where $D = r^{-1}d/dr(rd/dr)$ and $2m\varepsilon_F - k^2 \equiv k_F^2 \sin^2\alpha$. To be more precise, one has to solve the above equations simultaneously with the gap equation (3.95). We will be content ourselves, however, with solutions u, v^*, given in Fig. 5.5, which are obtained under the assumption that $\Delta(r)$ is proportional to $\psi(r)$.

Let us first note that Eq. (5.34) is left invariant under the substitutions $\varepsilon \to -\varepsilon$, $l \to -l+1$ and $u \to -v^*$. (All the negative-energy eigenstates are occupied at $T = 0$.) Therefore one may assume ε positive and study only the states with $l = 1$ since they give the lowest energy. One may assume $k_F^2 \sin^2\alpha \gg 2m\Delta$ since $\varepsilon_F \gg \Delta$. Then one finds from Eq. (5.34) that the approximate forms of the solution are $u(r) = f_u(r)J_1(k_F r \sin\alpha)$ and $v^*(r) = f_v(r)J_0(k_F r \sin\alpha)$, where the envelope functions f_u and f_v vary with the characteristic scale ξ. Here the indices of u and v^* are omitted. One expects that there are scattering states with $\varepsilon > \Delta_0$ and bound states with $\varepsilon < \Delta_0$. In fact, there is a bound state with the wave number $k_F \sin\alpha$ localised near the centre of the vortex with extension ξ, whose energy eigenvalue is $\varepsilon \sim (\pi/4)(\Delta_0^2/\varepsilon_F)g(\alpha)/\sin\alpha$, see [C-1]. Here $g(\alpha)$ is a slowly varying function of order unity. Thus there are almost gapless excitations localised near the vortex core and accordingly the density of states (5.33) varies spatially.

A single vortex has been considered so far. The density of states due to low energy excitations in a vortex lattice is larger along lines connecting the vortex with the nearest neighbour vortices. This has been confirmed by the STM images. It is interesting to note that a six-fold symmetry of a $NbSe_2$ crystal around the c-axis is also seen in the STM images of a vortex.

5.5 Time-dependent GL equation

Let us generalise the GL equation introduced in Section 5.1 so that we may study phenomena in which the order parameter Ψ depends on both space and time. That is, assuming that temporal variations are slow enough, one introduces a time derivative term in Eqs. (5.5) and (5.6). Unlike static cases, this extension is possible only under limited conditions. As is shown below, however, there are interesting applications of the time-dependent equation, the so-called TDGL equation.

The order parameter $\Delta(1)$ at a spacetir point $1 = (\boldsymbol{x}_1, t_1)$ is related to Δ at nearby points. (Here the BCS model is ⌣mployed and hence Δ is taken to be the order parameter.) When Δ does not change, namely when $\Delta = \Delta(1)$ at any nearby point, the rule by which Δ's are related is just the gap equation $\Delta(1) = gI(|\Delta(1)|)\Delta(1)$, where the gap in $I = N(0)\int d\xi \frac{1}{E}\tanh\frac{\beta E}{2}$ is $|\Delta(1)|$. If there is a variation $\delta\Delta(2) = \Delta(2) - \Delta(1)$ between the gaps at 2 and 1, we can express this relation by taking the effect of the variations into account, as

$$\Delta(1) = I\Delta(1) - g\int d2\{K_{11}(1,2)\delta\Delta(2) + K_{12}(1,2)\delta\Delta^*(2)\}, \tag{5.35}$$

where we have only kept terms linear in $\delta\Delta$, assuming the variation $\delta\Delta(2)$ is small. Causality restricts the t_2-integration within the range $t_2 \le t_1$. The second term of Eq. (5.35) represents the response to variations $\delta\Delta$ and $\delta\Delta^*$ of the mean fields. This term yields the linear response equation (3.48) with Q and C replaced by the mean-field variations $\delta\Delta$ and $\delta\Delta^*$.

It is sufficient to consider the case $\Delta(1) \propto e^{iq\cdot 1}$, that is, $\delta\Delta(2) = [e^{iq(2-1)} - 1]\,\Delta(1)$, for the present purpose, where $q = (\boldsymbol{q}, \omega)$. If this is the case, Eq. (5.35) becomes

$$[1 - I(|\Delta(1)|)]\Delta(1) = -g[K_{11}(q) - K_{11}(0)]\Delta(1) - g[K_{12}(q) - K_{12}(0)]\Delta^*(1). \tag{5.36}$$

In a superconductor without impurity scattering, one obtains

$$\begin{aligned} &K_{11}(q) \\ &= \sum_k \left\{ (1 - n_+ - n_-)\left(\frac{|u_+u_-|^2}{\omega + \varepsilon_+ + \varepsilon_-} - \frac{|v_+v_-|^2}{\omega - \varepsilon_+ - \varepsilon_-} \right) \right. \\ &\quad \left. -(n_+ - n_-)\left(\frac{|u_+|^2|v_-|^2}{\omega + \varepsilon_+ - \varepsilon_-} - \frac{|u_-|^2|v_+|^2}{\omega - \varepsilon_+ + \varepsilon_-} \right) \right\}, \end{aligned} \tag{5.37}$$

where the subscripts $\pm$ correspond to $\xi_k \pm v_F\boldsymbol{q}\cdot\boldsymbol{k}/k$ and ω is assumed to have a small imaginary part $+i\delta$. Provided that K may be expanded in powers of $\boldsymbol{q}^2$ and ω and after the replacements $\boldsymbol{q}^2 \to -\nabla^2$ and $\omega \to i\partial/\partial t$, Eq. (5.36) turns out to be the partial differential equation we are seeking. Although the denominator of the first term of Eq. (5.37) may be expanded in powers of ω when $\omega \ll 2\Delta$, the second term has a limit depending on the ratio qv_F/ω in the limit $qv_F \to 0, \omega \to 0$ and the above expansion is not then allowed.

For the case $T \sim T_c$ and $k_BT_c \gg v_Fq \gg |\Delta| \gg \omega$ the expansion is possible. If these conditions are satisfied, one obtains $K_{22} \propto \Delta^2$, which is thus negligible, and one simply considers the expression

$$K_{11} = \frac{N(0)}{2}\int_{-1}^{1} \mathrm{d}\mu \int \mathrm{d}\xi \tanh(\beta\xi/2)$$
$$\times[(2\xi + qv_{\mathrm{F}}\mu + \omega + \mathrm{i}\delta)^{-1} + (2\xi + qv_{\mathrm{F}}\mu - \omega - \mathrm{i}\delta)^{-1}] \quad (5.38)$$

in which Δ is set equal to 0. Since $v_{\mathrm{F}}q \gg \omega$, the imaginary part is dominant in the term proportional to ω in Eq. (5.38). Finally one obtains

$$K_{11}(\boldsymbol{q},\omega) - K_{11}(0,0) \simeq N(0)\left\{-\left(\frac{\beta_{\mathrm{c}}}{2\pi\mathrm{i}}\right)\frac{\pi^2}{6}\omega - \left(\frac{\beta_{\mathrm{c}}^2}{2}\right)\frac{7\zeta(3)}{8\pi^2}\frac{v_{\mathrm{F}}^2}{3}q^2\right\}. \quad (5.39)$$

The second term of the above equation is the gradient term obtained previously while the first term corresponds to the time derivative $-\tau(\partial\Psi/\partial t)$ in the GL equation. Comparing the first term with the gradient term, we determine the relaxation time τ

$$\tau = \frac{\pi}{8k_{\mathrm{B}}T_{\mathrm{c}}}. \quad (5.40)$$

This result is in agreement with the expectation that the order parameter relaxes towards a free energy minimum according to the equation

$$\tau\frac{\partial\Psi}{\partial t} = -\frac{\delta\tilde{F}}{\delta\Psi^*}, \quad (5.41)$$

where $\tilde{F}H_{\mathrm{c}}^2/4\pi$ is identical with F in Eq. (5.4). The relaxation mechanism is considered to be the scattering of thermally excited quasi-particles by the time-dependent mean field. It should be noted that τ remains unchanged while the coherence length ξ changes by non-magnetic impurity scattering.

As was noted in Section 3.6, the order parameter Ψ has a phase factor $\mathrm{e}^{-2\mathrm{i}\mu t}$ even in an equilibrium state, where μ is the chemical potential of an electron. Therefore, by noting that the chemical potential changes as $\mu \to \mu + e\phi$ in the presence of a scalar potential ϕ, one arrives at a correct expression by making the following replacement

$$\partial/\partial t \to \partial/\partial t + 2\mathrm{i}(\mu + e\phi).$$

This replacement makes the GL equation invariant under general gauge transformations. We thus rewrite Eq. (5.41) correctly as

$$\tau\left[\frac{\partial}{\partial t} + 2\mathrm{i}(\mu + e\phi)\right]\Psi + \xi^2\left(\frac{\nabla}{\mathrm{i}} - \frac{2\pi}{\phi_0}\boldsymbol{A}\right)^2\Psi - \Psi + |\Psi|^2\Psi = 0. \quad (5.42)$$

The TDGL equation of this simple diffusion type is valid only in a very narrow temperature region near T_{c} because of the group velocity anomaly of quasi-particles and the divergence of the density of states at the gap edge. In a gapless state due to magnetic impurity scattering studied in Section 4.6, the relaxation time τ is replaced by τ_p, in which case the theory has a wider applicability. This suggests that the dissipation term becomes larger if there are pair-breaking interactions. In reality, however, the contribution of inelastic scattering, due to an electron–phonon interaction, to the relaxation term is large and it is shown that the first term of Eq. (5.42) should be replaced by

$$\tau[1+(2\tau_E|\Delta|)^2]^{-1}\left[\frac{\partial}{\partial t}+2i(\mu+e\phi)+2\tau_E{}^2\frac{\partial|\Delta|^2}{\partial t}\right]\Psi. \tag{5.43}$$

Here τ_E is the relaxation time due to inelastic scattering and it is known that $2\tau_E\Delta_e$ takes a large value, where Δ_e is the mean field at an equilibrium state. For example the maximum value is 17 for Pb and 47 for Sn, see References [D-7] and [D-8] for details.

Another case where a differential equation for Ψ may be obtained is a system at $T \simeq 0$ with little thermal excitation, in which case such physical quantities as K in Eq. (5.37) are expanded in powers of ω/Δ_0 and qv_F/Δ_0. As a result, one obtains a wave equation, containing an operator $[\partial^2/\partial t^2-(v_F^2/3)\nabla^2]$, whose solutions include a Nambu–Goldstone mode with a spectrum $\omega = qv_F/\sqrt{3}$. However, time variation of the order parameter Ψ is in general accompanied by a density change $\delta\rho \propto i(\Psi^*\partial\Psi/\partial t-\Psi\partial\Psi^*/\partial t)$ as can be seen by studying the density response to change of pair mean field. Accordingly this mode is identified with plasma oscillation.

In the TDGL theory we assume, in addition to the restrictions already mentioned, that quasi-particle excitations are always in a local equilibrium state corresponding to $\Psi(x,t)$. Therefore there are some cases that defy description in terms of the TDGL equation. As was pointed out in Section 3.2, the charge density Q^* associated with quasi-particle excitations does not vanish if the system is in a nonequilibrium state so that the particle–hole symmetry is violated. Then the pair contribution Q_s deviates from the equilibrium value en_s and, as a result, there appears a state called the charge imbalance. An important example is a collective mode called the *Carlson–Goldman mode*. If there is a wave associated with the pair charge density Q_s, the charge Q^* oscillates out of phase to maintain charge neutrality. Although quasi-particle excitations decay by scattering, the above collective motion is a well-defined mode with a phase velocity $[(n_s/n)(4T/\pi\Delta)]^{1/2}v_F/\sqrt{3}$ near $T \simeq T_c$ since Q_s is small there. Although this mode is similar to the second sound in superfluid ^{4}He, where the superfluid and the normal fluid oscillate out of phase, the present mode is not an entropy mode; the restoring force of the wave motion comes from the change of the pair chemical potential, see [D-7, D-8] for details.

5.6 Applications of the TDGL equation

5.6.1 Sliding of vortex lattice

A vortex state appears in a type 2 superconductor when an external magnetic field H satisfies $H_{c1} < H < H_{c2}$, as was noted in Section 5.4. It is expected that the Lorentz force will act on the vortex if a current flows perpendicular to the flux. In fact, in an ideal system without vortex pinning centres, vortex lattice sliding appears, leading to energy dissipation caused by an applied electric field. Let us demonstrate this with the TDGL equation.

Suppose a scalar potential $\phi = -Ex$ giving a uniform electric field E is applied parallel to the x-axis, along with a vector potential $A = (0, Hx, 0)$. Then Eq. (5.42) becomes

$$\tau\left(\frac{\partial}{\partial t} - 2\mathrm{i}eEx\right)\Psi - \xi^2\partial^2\Psi/\partial x^2$$
$$-\xi^2(\partial/\partial y - \mathrm{i}2\pi Hx/\phi_0)^2\Psi - \Psi + |\Psi|^2\Psi = 0, \tag{5.44}$$

where physical quantities are equipped with proper dimensions for later convenience and the chemical potential μ has been included in the phase. Assuming that Ψ depends on time as $\Psi(x, y - vt)$, the first term of Eq. (5.44) becomes $-\tau v(\partial/\partial y + 2\mathrm{i}eEx/v)\Psi$. Accordingly if the choice $v = -cE/H$ is made so that the Hall voltage cancels the electric field E, this term and the third term may be combined. Therefore one obtains a vortex lattice solution sliding along the y-axis with velocity $v = cE/H$, if one superposes functions

$$\psi_p = \exp\left[\mathrm{i}p\left(y + \frac{cEt}{H}\right) - \frac{1}{2}r_0^2\left(\frac{x}{r_0^2} - p - \frac{\mathrm{i}\tau cE}{2H\xi^2}\right)^2\right]$$

as in Eq. (5.30) within the linear approximation introduced in Section 5.4.

Let us find the spatial average of the current density $j_x^{(\mathrm{L})}$ along the x-axis with the present solution. Since all ψ_p contain a phase factor $\exp(\mathrm{i}c\tau Ex/2\xi^2 H)$ proportional to x, one easily finds

$$\overline{j_x^{(\mathrm{L})}} = \frac{ec\tau}{2m\xi^2}\overline{|\phi|^2}\frac{E}{H}, \tag{5.45}$$

where $\overline{|\phi|^2}$ is the spatial average of the squared modulus of the order parameter and is related to the magnetisation through Eq. (5.42). Since E is finite, the total current density is obtained by adding Eq. (5.45) and $j_n = \sigma_n E$ when $H \sim H_{c2}$.

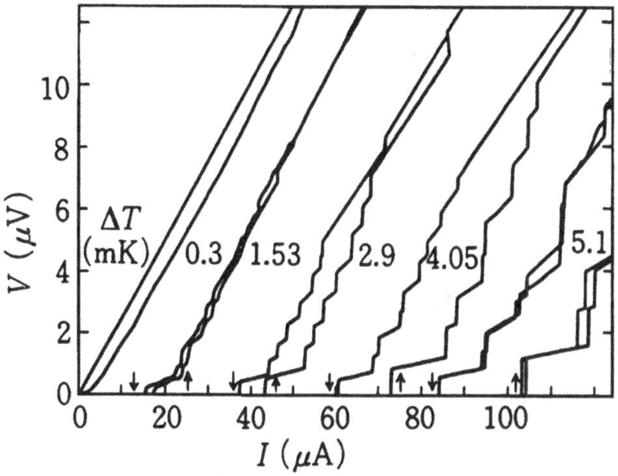

Fig. 5.9. The $I - V$ characteristic of a whisker of Sn (length 8×10^{-2} cm and cross section 1.93 μm^2). An arrow pointing up (down) denotes the data when a current is increased (decreased). $\Delta T = T_c - T$. (From [G-15].)

Even when the TDGL equation is not applicable, there is vortex lattice sliding, and hence energy dissipation, in an ideal system supporting a current perpendicular to a magnetic field H. In this respect, it is not adequate to call the state 'superconducting'. In reality, however, the vortex lines are pinned by various defects (*pinning*) and the supercurrent is stable in so far as the pinning force balances the Lorentz force. Superconducting magnets are possible because of this mechanism. Due to a low electron density or other reasons in the vicinity of a defect, the condensation energy of superconductivity is locally suppressed. The vortex energy is then lowered if the vortex core stays where the condensation energy is small. If the defect size d is larger than the coherence length, the energy is of the order of $H_c^2\xi^2 d/8\pi$. A distribution of these pinning centres will fix the whole vortex lattice since a vortex lattice has stiffness due to interactions between vortices. See [C-2, vol. II] and [E-6] for creep of lattices in the presence of pinning.

5.6.2 Phase slip oscillation (PSO)

The critical supercurrent density j_c of a thin wire has been obtained in Section 5.1. Related experiments have been carried out with a whisker crystal or a microbridge of such metals as Sn and In. Figure 5.9 shows an example of an $I - V$ characteristic thus obtained, from which one finds that the transition from a superconducting state ($V = 0$) to a normal state does not take place suddenly at $j = j_c$. Instead the voltage V increases stepwise

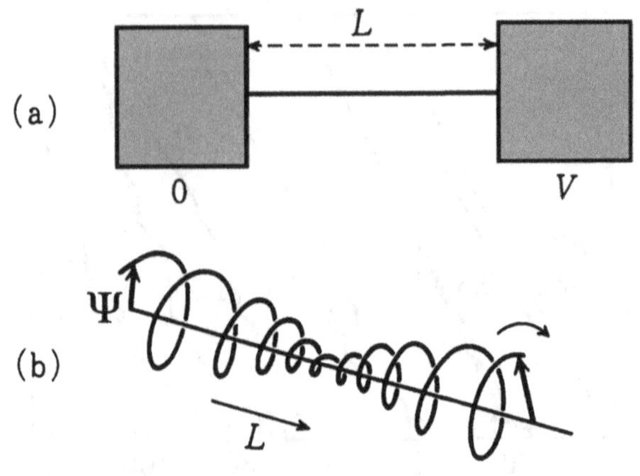

Fig. 5.10. Winding of the phase χ of Ψ in the presence of a potential difference V.

over a considerable range of applied current I. The first step is considered to be important and we will concentrate first on this phenomenon.

Suppose a thin wire of length L connects two large superconductors as shown in Fig. 5.10(a). If the electric field is uniform along the wire and the potential difference is finite, the phase of $\Psi = |\Psi|\,\mathrm{e}^{\mathrm{i}\chi}$ is given by $\chi(x,t) = -2eVxt/L$ and therefore the supercurrent increases in proportion to $v_s(x,t) = -eVt/mL$. Accordingly the system cannot be in a steady state. As v_s increases, $|\Psi|$ decreases and there appears a region where $|\Psi|$ is smaller than the average. The velocity v_s in this region then further increases, which leads to a further decrease of $|\Psi|$ and eventually $|\Psi|$ vanishes, see Fig. 5.10(b). The phase jumps by 2π at the same time, which implies that the phase difference created by the potential difference decreases by 2π. Then v_s becomes smaller and $|\Psi|$ is restored. Since the jump in the phase is 2π, $|\Psi|$ at the two sides of the zero point are smoothly connected. This unwinding mechanism repeats itself. If the phase jump takes place at intervals of τ_{ps}, the equality of the winding by $2eV$ per unit time and the unwinding described above leads to the Josephson relation $2eV = 2\pi/\tau_{ps}$, from which one obtains the average potential difference. This periodic phase jump is called phase slip oscillation (PSO). Early theories, which explained the $I - V$ characteristic near j_c using PSO, assumed the phase slip but later it was shown that PSO naturally arises as a solution to the TDGL equation.

Since there appears a point where $|\Psi|$ vanishes and a normal state sets in, to observe the charge neutrality, that is, the contribution j_n of the normal component to the current, one has to consider thermally excited quasi-particles. Note that a time-dependent electric field may exist in a

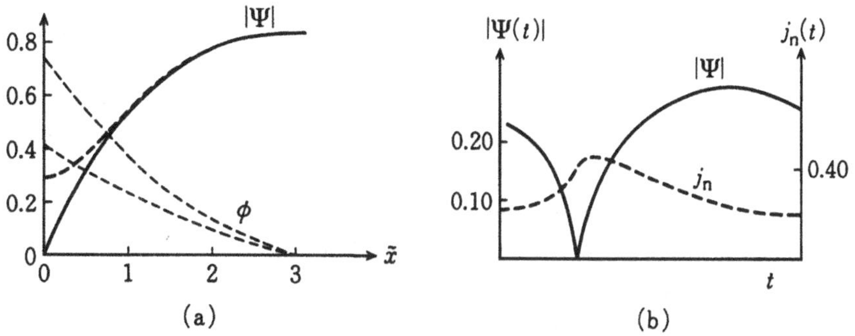

Fig. 5.11. (a) The spatial profiles of $|\Psi|$ and ϕ when $|\Psi(0,t)|$ takes a minimum and a maximum value in a PSO cycle. The coordinate $\tilde{x} = x/6d$ is measured from the phase slip centre. (b) The temporal variations of $|\Psi|$ and j_n. (From [G-16].)

superconductor although a static field does not. One can expect that j_n at $T \lesssim T_\mathrm{c}$ obeys the same Ohm's law $j_\mathrm{n} = \sigma_\mathrm{n} E$ as in the normal state. Since a quasi-one-dimensional system is under consideration, one may put $A = 0$ in j_s given by Eq. (5.7). Accordingly the total current is

$$j = \frac{en_\mathrm{s}}{2m}|\Psi|^2\frac{\partial \chi}{\partial x} - \sigma_\mathrm{n}\frac{\partial \phi}{\partial x}, \tag{5.46}$$

where $\chi(x,t)$ is the phase of Ψ and $\phi(x,t)$ denotes the scalar potential. The problem is to find solutions of the TDGL equation

$$\tau\left(\frac{\partial}{\partial t} + 2ie\phi\right)\Psi - \xi^2\frac{\partial^2\Psi}{\partial x^2} - \Psi + |\Psi|^2\Psi = 0, \tag{5.47}$$

under the condition that the current j given by Eq. (5.46) is constant. (It is sufficient to use Eq. (5.42) for qualitative discussions although Eq. (5.43) must be employed to compare the theory with experiments.)

One first notes that there are two uniform solutions: the normal state solution $|\Psi| = 0$, $\phi \propto x$; and the superconducting solution $|\Psi| = \mathrm{const.}$, $\chi \propto x$, $\phi = 0$. Although the first of these is stable for $j > j_\mathrm{c}$, there exists a third solution corresponding to PSO, which is more stable than the second solution in a certain region $j_l < j < j_\mathrm{c}$. This solution appears as a limit cycle when a relevant initial condition is chosen and, in practice, is analysed numerically. In an actual calculation, a periodic boundary condition over an interval $-d/2 < x \leq d/2$ is imposed and the point $x = 0$ is taken to be the *phase slip centre (PSC)* where the order parameter $|\Psi|$ vanishes periodically. Figure 5.11(a) shows the spatial profiles of $|\Psi|$ and ϕ at moments when $|\Psi(0,t)| = 0$ and $|\Psi(0,t)| = \max$. Figure 5.11(b) shows, on the other hand, the time dependences of $|\Psi|$ and j_n at the phase slip centre $x = 0$. In practice the

PSC appears at the most 'vulnerable' point on a thin wire. The jump in the $I - V$ characteristic is understood as a transition from the second solution with $\phi = 0$ to the PSO solution with $\phi \neq 0$. The successive appearance of steps as j is increased is attributed to the creation of many PSCs on the wire, see [D-8] for details.

Phase slip is also possible when the temperature is close to T_c, if points in a normal state appear due to thermal fluctuations. According to this mechanism, the resistance is proportional to the Boltzmann factor $e^{-\beta \Delta F}$, where $\Delta F \sim \xi S(H_c^2/8\pi)$, S being the cross section of the wire.

5.7 Fluctuations in a superconductor

We have studied the mean-field theory so far, in which only the statistical average of the order parameter Ψ is considered. This is, of course, an approximation and there are phenomena which cannot be understood within this framework. Among these, the most important is the critical phenomena near T_c in which order parameter fluctuations play a crucial rôle.

Let us restrict our discussion to superconducting fluctuations in a normal phase at $T > T_c$. Therefore $\Psi(x)$ appears only through thermal fluctuations and, as a result, a state with the order parameter profile $\Psi(x)$ is realised with the Boltzmann weight $\exp(-\beta F)$, where the free energy $F(\Psi(x))$ is given by Eq. (5.4). The contribution of thermal fluctuations to thermodynamic quantities is derived from the partition function

$$Z = \sum_{\{\Psi\}} \exp[-\beta F(\Psi(x))]. \tag{5.48}$$

Although F has been written in the above as if it were a functional of $\Psi(x)$ only, to be more precise, it also depends on the fluctuation δA of the gauge field A. The fluctuation δA is of the second order in Ψ, however, as can be seen from Eqs. (5.6) and (5.7), and accordingly the magnetic energy in F, induced by the fluctuation of Ψ, is of the fourth order in Ψ. Note also that the magnetic energy is proportional to κ^{-2} in a system with a large GL parameter ($\kappa \gg 1$). Therefore one may neglect δA in the Gaussian approximation, in which terms are kept up to the second order in Ψ, or in a system with $\kappa \gg 1$.

Let us consider, as the simplest example, fluctuations in a uniform system without an external magnetic field. Since the dependence of physical quantities on $\varepsilon \equiv T/T_c - 1$ is most important in a critical phenomenon, one writes

$$\beta F = \tilde{\beta} \int \mathrm{d}\boldsymbol{x} \left\{ {\xi_0}^2 \nabla \Psi^* \nabla \Psi + \varepsilon |\Psi|^2 + \frac{1}{2} |\Psi|^4 \right\}$$
$$= \tilde{\beta} \sum_k ({\xi_0}^2 k^2 + \varepsilon) {\psi_k}^* \psi_k + \text{(fourth order terms)}, \qquad (5.49)$$

where $\tilde{\beta} = \beta H_c^2(T)/4\pi\varepsilon$, $\xi_0^2 = \varepsilon\xi^2$ and the Fourier transform of $\Psi(\boldsymbol{x})$ is denoted by $\psi_k = \psi_{1k} + \mathrm{i}\psi_{2k}$, where ψ_{1k} and ψ_{2k} are independent fluctuation variables for each $\boldsymbol{k}$. The fluctuation amplitude is small, unless ε is very small, and thus the Gaussian approximation which neglects $|\Psi|^4$-terms is applicable. Then the statistical average of the fluctuation is obtained from

$$Z = \prod_k \iint_{-\infty}^{\infty} \mathrm{d}\psi_{1k} \mathrm{d}\psi_{2k} \exp\{-\tilde{\beta}({\xi_0}^2 k^2 + \varepsilon) \times ({\psi_{1k}}^2 + {\psi_{2k}}^2)\}$$
$$= \prod_k \pi \tilde{\beta}^{-1} ({\xi_0}^2 k^2 + \varepsilon)^{-1}. \qquad (5.50)$$

Among contributions to the specific heat, for example, the term which diverges as $\varepsilon \to 0$ is

$$C_\mathrm{F} = k_\mathrm{B} \sum_k ({\xi_0}^2 k^2 + \varepsilon)^{-2} \propto (k_\mathrm{B}/{\xi_0}^d) \varepsilon^{(d-4)/2}, \qquad (5.51)$$

where d is the dimension of the system. Therefore C_F diverges as $\varepsilon^{-1/2}$ as $T \to T_\mathrm{c}$ in a three-dimensional sample. Let us recall that in the mean-field approximation there appears merely a jump in the specific heat at $T = T_\mathrm{c}$. The coefficient in front of ε^{-1} is, however, very small compared to the electron specific heat $N(0)k_\mathrm{B}^2 T_\mathrm{c}$ in the normal state and hence C_F may be negligible. However the contribution of this term to the electrical conductivity is sizeable as will be shown below. We note that

$$\langle \psi_k {\psi_{k'}}^* \rangle = \delta_{kk'} \tilde{\beta}^{-1} ({\xi_0}^2 k^2 + \varepsilon)^{-1} \qquad (5.52)$$

in the same approximation, from which the correlation function for Ψ is obtained as

$$\langle \Psi(x) \Psi^*(x') \rangle = \left[4\pi\tilde{\beta}{\xi_0}^2 |\boldsymbol{x} - \boldsymbol{x}'| \right]^{-1} \exp(-|\boldsymbol{x} - \boldsymbol{x}'|/\xi). \qquad (5.53)$$

This shows that the correlation length of the order parameter diverges as ξ, that is, as $\varepsilon^{-1/2}$ as $T \to T_\mathrm{c}$.

5.7.1 Electrical conductivity due to fluctuations

Since a local superconducting state is created and annihilated continually even at $T > T_\mathrm{c}$, these fluctuations are expected to manifest themselves in the electrical conductivity. It is convenient to employ the Kubo formula in the classical approximation to calculate this quantity. This corresponds to the imaginary part proportional to ω in the approximate form of Eq. (3.48) with $\beta\omega \ll 1$. One thus obtains

$$\sigma_{xx}(\boldsymbol{q},\omega) = \beta \int_0^\infty \langle j_x(\boldsymbol{q},t) j_x(\boldsymbol{q},0)\rangle \cos\omega t \, \mathrm{d}t, \tag{5.54}$$

where

$$j_x(\boldsymbol{q},t) = \frac{en_s}{4m} \sum_k (2k_x + q_x)\psi^*_{k+q}(t)\psi_k(t), \tag{5.55}$$

since the external field is set equal to zero. The symbol $\langle\cdots\rangle$ denotes the statistical average with respect to the fluctuations. One has to know the time dependence of ψ_k, namely how the fluctuation $\Psi(\boldsymbol{x})$ relaxes to 0, to evaluate Eq. (5.54). A frequently employed assumption is that $\Psi(\boldsymbol{x},t)$ obeys the simple TDGL equation (5.42), which we also assume here. Accordingly ψ_k in the Gaussian approximation becomes

$$\psi_k(t) = \psi_k(0) \exp\left[-({\xi_0}^2 k^2 + \varepsilon)t/\tau_0\right], \tag{5.56}$$

where $\tau_0 = \pi/8k_B T_c$ in the absence of impurity scattering ($l \gg \xi_0$). The conductivity $\sigma(q,\omega)$ is found if Eqs. (5.55) and (5.56) are substituted into Eq. (5.54) and then $\langle|\psi_k|^4\rangle = 2\tilde{\beta}^{-2}(\xi_0^2 k^2 + \varepsilon)^{-2}$ is used. The direct current conductivity $\sigma(q = 0, \omega = 0)$ is found to be

$$\sigma = {\beta_c}^{-1}(e{\xi_0}^2)^2 \frac{\tau_0}{2} \frac{1}{d} \sum_k k^2 \frac{1}{({\xi_0}^2 k^2 + \varepsilon)^3}, \tag{5.57}$$

where d stands for the dimensionality of the system and the relation $n_s/2m = H_c^2 \xi^2/\pi$ has been used. The result of the integration yields

$$\sigma_3 = \frac{e^2}{32\hbar} \cdot \frac{1}{\xi_0 \varepsilon^{1/2}} \tag{5.58a}$$

$$l \cdot \sigma_2 = \frac{e^2}{16\hbar} \cdot \frac{1}{\varepsilon} \tag{5.58b}$$

$$S \cdot \sigma_1 = \frac{e^2}{16\hbar} \cdot \frac{1}{\varepsilon^{3/2}}. \tag{5.58c}$$

In the above equation, l is the thickness of a thin film while S is the cross section of a thin wire and it is assumed that $l, \sqrt{S} \ll \xi_0$. It is interesting to note that the conductance in Eqs. (5.58b, c) is expressed in terms of the elementary unit $e^2/\hbar$ only.

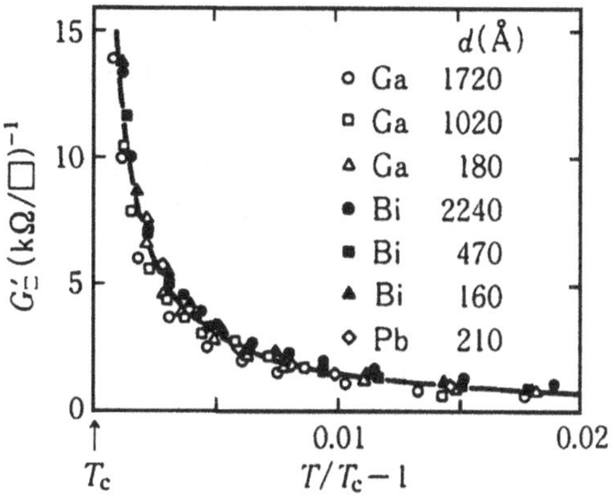

Fig. 5.12. The conductance due to fluctuations in a thin film of 1 cm × 1 cm with thickness *d*. The solid line is the prediction by the Aslamasov–Larkin theory. (From [E-4].)

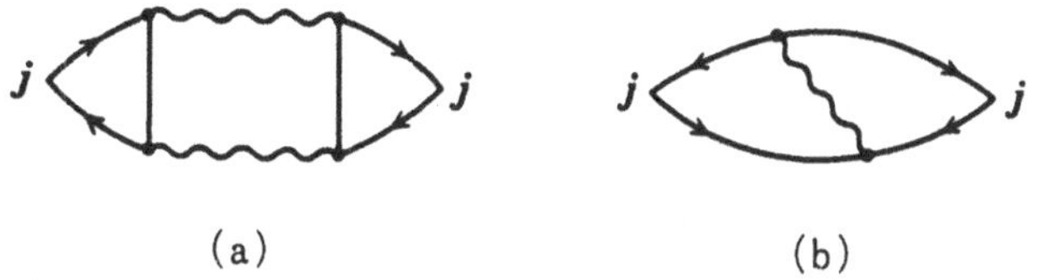

Fig. 5.13. Contribution of fluctuations (wavy lines) to conductance σ. (a) The AL term, and (b) the MT term.

σ_3 is far smaller than the normal state conductivity $\sigma_n \sim ne^2\tau/m \sim k_F^2 le^2$ and is difficult to observe (see Chapter 7 for high-T_c superconductors). On the other hand, sheet conductance measurement in a thin film shows an excellent agreement between Eq. (5.58b) and experiment, see Fig. 5.12.

In fact, the results obtained above take into account only a part of the fluctuations of Ψ, that is, Fig. 5.13(a) in the graphical representation of the $j - j$ correlation (5.54). This so-called Aslamasov–Larkin (AL) term should be supplemented by the process in Fig. 5.13(b), which is important in quantitative discussions of electrical conductivity at $T \lesssim T_c$ and is called the Maki–Thompson (MT) term.

5.7.2 Ginzburg criterion

Thus far, we have employed the Gaussian approximation where the $|\Psi|^4$-term in F has been neglected. As the temperature T approaches T_c and ε becomes small, however, fluctuations grow so large that this approximation is no longer applicable. Then the system is in the critical region where thermodynamic quantities blow up with anomalous powers of ε. The order of magnitude of ε under which the system is in the critical region is called the Ginzburg criterion.

Let us consider the averaged order parameter $\bar{\Psi} = \frac{1}{V}\int_V \Psi(x)\mathrm{d}x$ over a volume $V \sim \xi^3$ since typical fluctuations have a linear size $\sim \xi$. The statistical average of the modulus of $\bar{\Psi}$ is obtained from

$$\begin{aligned}\langle \bar{\Psi}^* \bar{\Psi} \rangle &= \frac{1}{V} \int \langle \Psi^*(x)\Psi(0) \rangle \, \mathrm{d}x \\ &\sim \frac{1}{\xi^3} \frac{1}{\tilde{\beta}{\xi_0}^2} \xi^2 = \frac{1}{\tilde{\beta}{\xi_0}^3} \varepsilon^{1/2},\end{aligned}$$

where Eq. (5.53) has been used. When the above average becomes of the order of ε, one finds that the fourth order term $\langle |\bar{\Psi}|^4 \rangle \sim \langle |\bar{\Psi}|^2 \rangle^2$ is of the order of $\varepsilon \langle |\bar{\Psi}|^2 \rangle$, which indicates that this term is not negligible. Thus it follows that the condition for the system to be in the critical region is $(\tilde{\beta}\xi_0^3)^{-2} \gtrsim \varepsilon$, which may be rewritten as

$$(\varepsilon_\mathrm{F}/k_\mathrm{B} T_\mathrm{c})^2 (n{\xi_0}^3)^{-2} \gtrsim \varepsilon. \tag{5.59}$$

For an ordinary superconductor, the left hand side is less than 10^{-12} and the critical region is negligibly small. ε is of the order of 10^{-2} even for a high-T_c superconductor for which ξ_0 is small. A different situation appears in a high magnetic field as will be discussed in Chapter 7.

It should be remarked that fluctuations are quite important in superfluid ^{4}He in contrast with superconductors. This is for two reasons: the length ξ defined by Eq. (1.9), which corresponds to the coherence length ξ_0, is of the order of the interparticle distance; and since the condensation energy for one particle is $\sim k_\mathrm{B} T_\mathrm{c}$, the left hand side of Eq. (5.59) is

$$(n{\xi_0}^3)^{-2} \sim 1. \tag{5.60}$$

Accordingly properties near the critical point deviate significantly from the mean-field predictions. A good example of this is the temperature dependence of the specific heat C shown in Fig. 1.1(b), where C is observed to diverge remarkably at the critical point due to critical fluctuations.

6

Superfluid ^{3}He

Superfluidity in liquid ^{3}He, discovered in 1972, is due to pairing, as is the case for superconductivity in ordinary metals. However, the pair is in the ^{3}P state so that a new aspect of superfluidity appears with internal degrees of freedom. It offers quite an interesting real example of 'spontaneous symmetry breaking'. This aspect will be emphasised in the following discussion.

6.1 Fermi liquid ^{3}He

A ^{3}He atom is a fermion with nuclear spin 1/2. Although ^{3}He gas liquefies at low temperatures due to the weak van der Waals force, it does not solidify even at $T = 0$ unless the pressure exceeds 33.5 atm (3.4×10^6 Pa) as shown in the phase diagram (Fig. 6.1(a)). While Bose liquid ^{4}He becomes superfluid at about 2 K, ^{3}He is Fermi-degenerate at less than $T_{\mathrm{F}} \sim 0.1$ K and shows the properties of a normal Fermi gas. For example, the specific heat and the susceptibility are proportional to T at $T < T_{\mathrm{F}}$. It is only at an ultra-low temperature in the mK-region that the system undergoes a transition to the superfluid state.

Figure 6.1(b) is the phase diagram showing the superfluid phases with variables T, P and external magnetic field H in the mK-region. It should be noted that there are two thermodynamically distinct phases called the A phase and the B phase even at $H = 0$, and that for $H \neq 0$ the A_1 phase appears. Moreover, the transition between the A phase and the B phase is of the first order, which suggests that these are ordered phases with different symmetries. The appearance of superfluidity has been suggested by the specific heat jump, observation of the fourth sound propagating in the 'super leak' of packed fine powder, the anomaly in heat conductance and, above all, the shift in nuclear magnetic resonance frequency (see Section 6.4).

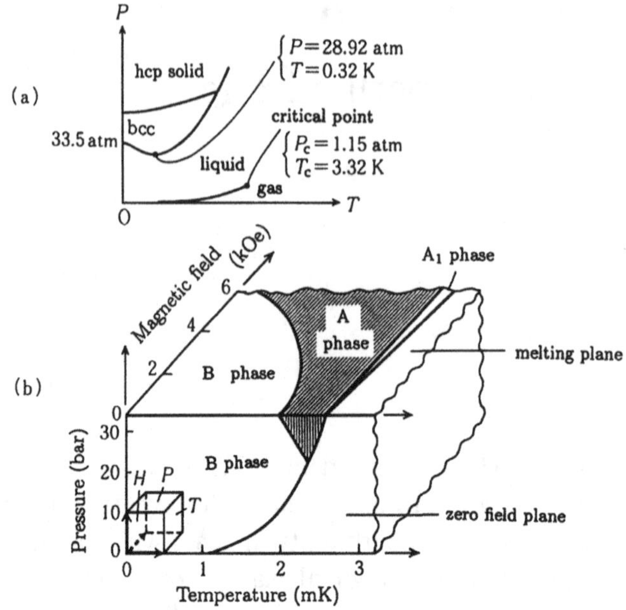

Fig. 6.1. (a) The phase diagram of ^{3}He at low temperatures. (b) The phase diagram of liquid ^{3}He in the mK-region. There appear three distinct superfluid phases denoted by A, A_1 and B.

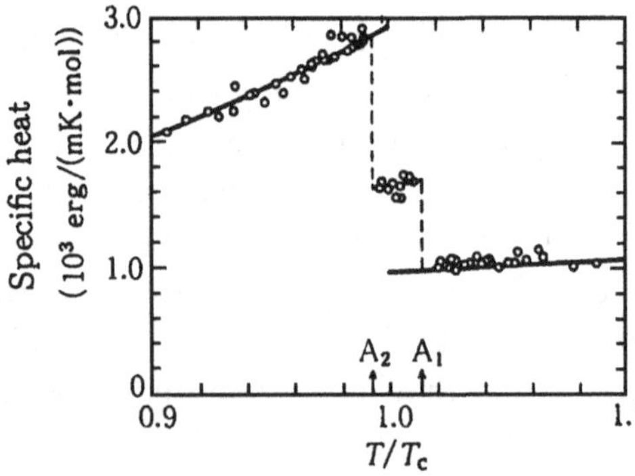

Fig. 6.2. Temperature dependence of the specific heat with a magnetic field 0.88 T under melting pressure. Jumps are seen in the transition from the normal phase to the A_1 phase (A_1) and that from the A_1 phase to the A phase (A_2). (From [G-17].)

The existence of the A_1 phase in a magnetic field is deduced from the specific heat measurement (Fig. 6.2) for example.

The starting point of the theory of the superfluid state in liquid ^{3}He is, as in the case of superconductivity, the model for the normal state. In addition to the van der Waals force, there is a hard core-like repulsive potential between ^{3}He atoms and, as a consequence, the interaction can not be considered weak. This should be evident if one compares the interatomic distance 2.88×10^{-8} cm, at which the repulsion starts to operate, and the mean interparticle distance (under saturated vapour pressure) $\bar{r} = 3.49 \times 10^{-8}$ cm in the liquid state. In fact, liquid ^{3}He undergoes a phase transition to a bcc crystal due to this repulsive force if the particle density is increased by 5% under applied pressure. It should also be noted that the specific heat and the magnetisation calculated within the free particle model greatly deviate from the observed values. In spite of these facts, the lifetime of a quasi-particle excitation dressed by interactions is rather long owing to the surrounding degenerate Fermi sea. Accordingly the physical properties at $T \ll T_{\rm F}$ may be well described by a theory based on quasi-particle excitations, the so-called Landau Fermi liquid theory, see also [A-6]. Since the critical temperature $T_{\rm c}$ of this superfluid is of the order of $T_{\rm F} \times 10^{-3}$, similar to that for superconductivity in a metal, one can naturally study the system on the basis of this theory.

In the Fermi liquid theory, the quasi-particle is also specified by the momentum $\boldsymbol{p}$ and the spin α, hence being in one-to-one correspondence with the original particle. The normal ground state of the system is represented by the Fermi sphere with radius $k_{\rm F} = (3\pi^2 n)^{1/3}$ and an excited state is specified by the deviation $\delta n_{p\alpha}$ of the quasi-particle distribution from the Fermi sphere. It is assumed, furthermore, that not only direct interaction but interaction via surrounding particles (i.e., polarisation of the surrounding particles) acts between excited quasi-particles. The effective interaction is customarily separated into spin-independent and spin-dependent parts as

$$H_{\rm int} = \frac{1}{2}\sum\{V^{(\rm s)}(\boldsymbol{p},\boldsymbol{p}',\boldsymbol{q})a^\dagger_{p+q,\alpha}a_{p\alpha}a^\dagger_{p'\beta}a_{p'+q,\beta} + V^{(\rm a)}(\boldsymbol{p},\boldsymbol{p}',\boldsymbol{q})a^\dagger_{p+q,\alpha}\boldsymbol{\sigma}_{\alpha\beta}a_{p\beta}a^\dagger_{p'\gamma}\boldsymbol{\sigma}_{\gamma\delta}a_{p'+q,\delta}\}. \tag{6.1}$$

To be more precise, $V^{(\rm s)}$ and $V^{(\rm a)}$ are the irreducible vertex parts of the channel indicated by an arrow in Fig. 6.3(a) and given by the scattering amplitudes of quasi-particles on the Fermi surface.

Equilibrium properties including the linear response are obtained from the generalised mean-field approximation for the effective interaction above. Thus, one uses the Hamiltonian

$$\mathscr{H}_{\rm mf} = \sum_{p,\alpha}\xi_{p\alpha}a^\dagger_{p\alpha}a_{p\alpha} + \frac{1}{2N(0)}\sum_{p,p'}\sum_{l=0}^{\infty}P_l(\hat{\boldsymbol{p}},\hat{\boldsymbol{p}}')\{F_l^{(\rm s)}a^\dagger_{p\alpha}a_{p\alpha}\delta\langle a^\dagger_{p'\beta}a_{p'\beta}\rangle + F_l^{(\rm a)}{\rm tr}\,(a^\dagger_p\boldsymbol{\sigma}a_p)\delta\langle {\rm tr}\,a^\dagger_{p'}\boldsymbol{\sigma}a_{p'}\rangle\} + \text{(interaction with external fields)}, \tag{6.2}$$

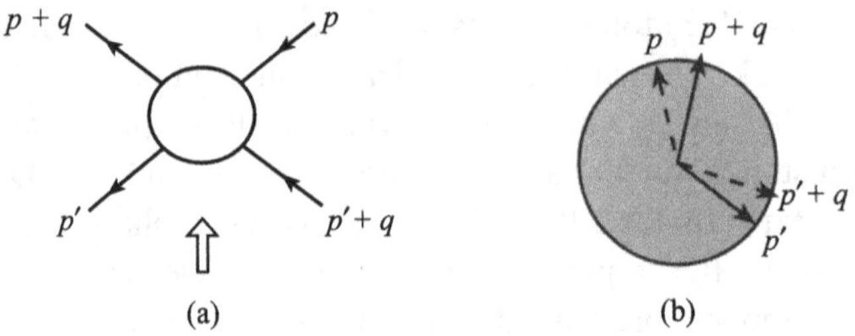

Fig. 6.3. Interparticle interaction.

where $\xi_{p\alpha} = p^2/2m^* - \mu$. The symbol $\delta\langle\cdots\rangle$ represents the deviation of a physical quantity from the mean value due to external fields, for example, in an equilibrium state and is a function of momentum and spin. One usually considers the quasi-particle density, the spin density or the momentum density as the physical quantity of interest. The ordinary Hartree–Fock term independent of external fields is considered to be renormalised into the chemical potential μ. The quantities $F_l^{(s)}$ and $F_l^{(a)}$ made dimensionless by $2N(0)$ are called the *Landau parameters.*

Suppose the chemical potential is changed by $\delta\mu$. The particle density then changes by $\delta n = \sum_{p,\alpha} \delta\langle n_{p,\alpha}\rangle$, which leads to a quasi-particle energy

$$\bar{\xi}_{p\alpha} = \xi_{p\alpha} - \delta\mu + F_0^{(s)}\delta n/2N(0). \tag{6.3}$$

Accordingly it follows that

$$\delta\langle n_{p\alpha}\rangle = \frac{\partial f_p^0}{\partial \xi_{p\alpha}}\left(-\delta\mu + \frac{1}{2N(0)}F_0^{(s)}\delta n\right),$$

where f_p^0 is the Fermi distribution function and $\partial f_p^0/\partial\xi_{p\alpha}$ may be approximated by a δ-function localised at the Fermi surface. Thus it follows immediately from

$$\delta n = \sum_{p,\alpha} \frac{\partial f_p^0}{\partial \xi_{p\alpha}}\left(-\delta\mu + \frac{1}{2N(0)}F_0^{(s)}\delta n\right) = 2N(0)\delta\mu - F_0^{(s)}\delta n \tag{6.4}$$

that the compressibility is given by

$$\kappa = \frac{\kappa_0}{1 + F_0^{(s)}}, \tag{6.5}$$

where $\kappa_0 = 2N(0)n^{-2}$ is the compressibility of a free Fermi gas. As is seen from Table 6.1, $F_0^{(s)}$ takes a large value due to the pressure from the hard core, suppressing density fluctuation.

Table 6.1. *The molar volume, the Landau parameters at $T < T_c$ and the critical temperature of liquid ^{3}He*

P (bar)	V (cm^3)	$n \times 10^{21}$ (cm^{-3})	$k_F \times 10^7$ (cm^{-1})	m^*/m	$F_1^{(s)}$	$F_0^{(s)}$	$F_0^{(a)}$	T_c (mK)	T_{AB} (mK)
0	36.84	16.3	7.84	2.80	5.39	9.30	−0.6951	0.929	–
15	28.89	20.8	8.49	4.28	9.85	41.73	−0.753	2.067	–
34.4	25.50	23.6	8.87	5.85	14.56	88.47	−0.753	2.491	1.933

Similarly, if a uniform magnetic field $\boldsymbol{B}$ ($\parallel \hat{z}$) is applied, one obtains

$$\delta\langle s_z\rangle = \sum_{p,\alpha} \frac{\partial f^0_{p\alpha}}{\partial \xi_{p\alpha}} \left[-\frac{1}{2}\gamma B + \frac{1}{2N(0)} F_0^{(a)} \delta\langle s_z\rangle \right], \tag{6.6}$$

corresponding to Eq. (6.4), where $\delta\langle s_z\rangle = \sum_{p,\alpha} \alpha\delta\langle n_{p\alpha}\rangle$ and $\gamma/2$ is the magnetic moment of a ^{3}He atom. Therefore the susceptibility is

$$\chi = \frac{\chi_n{}^0}{1 + F_0^{(s)}}, \tag{6.7}$$

which shows that the correction to $\chi_n^0 = \gamma^2 N(0)/2$ is given by $F_0^{(a)}$. The parameter $F_0^{(a)}$ is negative and its magnitude is rather large. Thus it is found that liquid ^{3}He is easily spin-polarised which is also due to the hard core repulsion.

Suppose finally that one references to a frame moving with a constant velocity $-\boldsymbol{v}$. The quasi-particle energy becomes $\xi_{p\alpha} + \boldsymbol{p} \cdot \boldsymbol{v}$ under the Galilei transformation. One only needs to replace $\delta\mu$ above by $\boldsymbol{p} \cdot \boldsymbol{v}$ to recover the original Fermi sphere from the Fermi sphere in the moving frame. That is,

$$\delta n_{p\alpha} = \frac{\partial f^0_p}{\partial \xi_{p\alpha}} \left\{ -\boldsymbol{p} \cdot \boldsymbol{v} + \frac{1}{N(0)} \hat{\boldsymbol{p}} \cdot \hat{\boldsymbol{p}}' F_1^{(s)} \delta n_{p'} \right\}.$$

One obtains, by making use of the above equation, an expression

$$\boldsymbol{J} = m^* n \boldsymbol{v} \left(1 + \frac{1}{3} F_1^{(s)}\right)^{-1}$$

for the current $\boldsymbol{J} = \sum_{p,\alpha} \boldsymbol{p}\delta n_{p\alpha}$. On the other hand, one should have $\boldsymbol{J} = nm\boldsymbol{v}$ since all the particles move with velocity $\boldsymbol{v}$ if seen from the moving frame. Therefore the effective mass is related to $F_1^{(s)}$ as

$$\frac{m^*}{m} = \left(1 + \frac{1}{3} F_1^{(s)}\right). \tag{6.8}$$

The parameter $F_1^{(s)}$ also appears in the expression for the specific heat. The values of these Landau parameters are listed in Table 6.1, from which one concludes that the interaction is by no means negligible.

6.1.1 Pairing interaction

It can be seen from Eq. (6.2) that Landau parameters give the magnitude of the effective interaction when $q \simeq 0$ in Eq. (6.1). This corresponds to the forward scattering amplitude with a small exchange momentum q from the viewpoint of a two-quasi-particle scattering, see Fig. 6.3(b). In contrast with this, it is that part of an interaction with rather large q, such that $q \lesssim 2k_F$, that leads to superfluid pairing. Therefore it is not easy to estimate the pairing interaction qualitatively from the magnitude of the parameters $F_1^{(s)}$ and $F_l^{(a)}$. (See [B-7] for numerous attempts to do this.) In qualitative discussions below we keep the above points in mind.

It was shown in Section 2.1 that an interaction can be separated into $V^{(e)}$ acting in spin-singlet pairs and $V^{(o)}$ acting in spin-triplet pairs in the absence of spin–orbit coupling. This decomposition and that into symmetric and antisymmetric parts $V^{(s)}$ and $V^{(a)}$ in Eq. (6.1) are related as follows. Since the projection operators onto a singlet-pair and a triplet-pair are $P_1 = (1 - \boldsymbol{\sigma} \cdot \boldsymbol{\sigma}')/4$ and $P_3 = (3 + \boldsymbol{\sigma} \cdot \boldsymbol{\sigma}')/4$ respectively, one has $P_1 + P_3 = 1$ and $\boldsymbol{\sigma} \cdot \boldsymbol{\sigma}' = P_3 - 3P_1$. If these relations are used in Eq. (6.1), one obtains

$$\begin{aligned} V^{(e)} &= V^{(s)} - 3V^{(a)} \\ V^{(o)} &= V^{(s)} + V^{(a)}. \end{aligned} \tag{6.9}$$

It is expected that the dominant contribution to pairing is provided by the component $l = 0$. (The angle with which the Landau parameters are decomposed into partial waves is of course different from that used in the decomposition of the pairing interaction into the s-wave, p-wave and so on.) We now note that $F_1^{(a)}$ takes a large negative value. It follows from Eq. (6.9) that if $V^{(a)}$ is negative, it works as a repulsive force for a spin-singlet pair and an attractive force for a spin-triplet pair. One may then presume that this is related to the fact that the triplet p-wave pairing is responsible for the superfluidity in ^{3}He. Conversely singlet s-wave pairing cannot be expected since $F_0^{(s)}$ takes a large positive value. In fact, right after the BCS theory was expounded there were theoretical predictions on the nature of the pairing in ^{3}He and leading opinion favoured either d-wave or p-wave pairing. It is established nowadays that the spin-triplet p-wave pairing is responsible for superfluidity in ^{3}He as is shown below. The problem of the interaction is considered again in Section 6.4 in the context of the strong coupling effect.

6.2 Superfluid state of ^{3}He pairs

The weak coupling theory for the superconducting state with ^{1}S pairs has been studied in Chapter 3. Let us repeat the same analysis here for ^{3}P pairs. It has already been shown that the order parameter for a spin-triplet pairing is expressed as $\Psi^{(3)} = A_\mu(\boldsymbol{k})\mathrm{i}\sigma_\mu\sigma_2$, where $A_\mu(\boldsymbol{k})$ transforms as a vector under a spin-space rotation. Let us assume here that the $l = 1$ component (i.e. p-wave) is the most attractive force when the interaction is decomposed as in Eq. (2.12). Therefore one has

$$V(\boldsymbol{k},\boldsymbol{k}') = -4\pi g_1 \sum_{m=-1}^{1} Y_1{}^m(\hat{\boldsymbol{k}})Y_1{}^{-m}(\hat{\boldsymbol{k}}')$$
$$= -3g_1(\hat{k}_x\hat{k}'_x + \hat{k}_y\hat{k}'_y + \hat{k}_z\hat{k}'_z), \tag{6.10}$$

where $\hat{\boldsymbol{k}} = \boldsymbol{k}/|\boldsymbol{k}|$, $Y_1^{\pm 1} = \sqrt{3/8\pi}(\hat{k}_x \pm \mathrm{i}\hat{k}_y)$, $Y_1^0 = \sqrt{3/4\pi}\hat{k}_z$ and g_1 stands for $|V_1|$. The order parameter $A_\mu(\boldsymbol{k})$ may be considered to be a function of the direction of $\boldsymbol{k}$ only and expanded in terms of the p-wave orbits $Y_1^m(\hat{\boldsymbol{k}})$ or $\hat{k}_j$ $(j = x, y, z)$ as

$$A_\mu(\hat{\boldsymbol{k}}) = \sum_{m=-1}^{1} A_{\mu m} Y_1{}^m(\hat{\boldsymbol{k}}) = \sum_j A_{\mu j}\hat{k}_j \tag{6.11}$$

for the interaction (6.1). Representation in terms of $\hat{k}_j$ will be mainly used in the following. Clearly the triplet $A_{\mu j}$ $(j = x, y, x)$ with a fixed index μ transforms as a vector under a rotation in momentum space. Accordingly the order parameter of a superfluid state with ^{3}P pairing is given by 3×3 $A_{\mu j}$'s with the spin component μ and the p-orbital component j. Moreover each $A_{\mu j}$ is a complex number since it is a pair wave function. In contrast with a ^{1}S pairing, for which the order parameter is a single complex number A, it is expected from the above arguments that the superfluidity with a ^{3}P pair can have far more diverse phases.

Given an ordered state represented by $\{A_{\mu j}\}$, it transforms to a different ordered state under the above mentioned operations, namely a spin-space rotation, a rotation in $\hat{\boldsymbol{k}}$-space and multiplication by a common phase $\mathrm{e}^{\mathrm{i}\chi}$ $(0 \le \chi < 2\pi)$ called a gauge transformation. The normal state is invariant under these transformations since the Hamiltonian does not contain the spin–orbit interaction. That the system is in an ordered state implies, therefore, the breaking of the symmetries these operations represent. Since the broken symmetries are rotations in two three-dimensional spaces and change in phase, the symmetry under consideration in the superfluid phase in ^{3}He is

$$G = SO_3 \times SO_3 \times U(1) \times T \tag{6.12}$$

in terms of the group theoretical symbols. Here the time-reversal operation T is also included. The action of T on the order parameter of a triplet pairing is

$$TA_\mu(\boldsymbol{k}) = A_\mu^*(-\boldsymbol{k}).$$

It should be noted that the symmetry operations in Eq. (6.12) are not mutually independent.

All the states of liquid ^{3}He represented by Eq. (6.11) share the common critical temperature T_c as was mentioned in Chapter 2. There is only a single class at T_c in this sense. Strictly speaking, there is a magnetic dipole interaction H_D, to be introduced in Section 6.4, which is a form of spin–orbit interaction, and the Hamiltonian is not invariant under separate rotations in momentum space and spin space. Accordingly different states have different T_c if H_D is taken into account. This splitting of T_c is, however, so small ($\sim 10^{-7}$ K) that it has not yet been observed and this effect is neglected in the present and the next sections. In neutron matter, due to a strong tensor force, the $J = 2$ pairing state, ^{3}P$_2$ superfluid in particular, is thought to have the highest T_c.

Liquid ^{3}He is classified into many different states at $T < T_\text{c}$. The symmetry G is not completely broken in an ordered state in general and some symmetry corresponding to a subgroup H of G may be left intact. In other words, it is the symmetry of the coset group G/H that is broken in many cases. The set of states constructed by operating G/H on one of these states form a class. Naturally all the states belonging to a certain class have the same free energy, although the free energies of states in different classes are not degenerate in general. While there is only one class in the ^{1}S pairing, each member of which is specified by the phase, there are many classes with different symmetries in the ^{3}P pairing as will be shown below. Since the free energies of these classes change with temperature, pressure and other parameters, there are possibilities that phase transitions between superfluid states take place and indeed these transitions are observed. Classes of typical ordered states, including those actually appearing in the superfluid phase of liquid ^{3}He, are given below. We note that related subjects will appear at the end of Chapter 7.

6.2.1 BW state

First consider the simplest form:

$$A_{\mu j} = A\delta_{\mu j}. \tag{6.13}$$

Written in the 2 × 2 matrix form this state is

$$\hat{\Psi} = A\begin{pmatrix} -\hat{k}_x + \mathrm{i}\hat{k}_y & \hat{k}_z \\ \hat{k}_z & \hat{k}_x + \mathrm{i}\hat{k}_y \end{pmatrix}. \tag{6.14}$$

The factor A contains a phase χ, $A = |A|\mathrm{e}^{\mathrm{i}\chi}$. This order parameter corresponds to a state with $J = |\boldsymbol{L} + \boldsymbol{S}| = 0$, that is, a state with condensation of $^3\mathrm{P}_0$ pairs. If this state is rotated round a direction $\hat{\boldsymbol{n}}$ by an angle θ in spin space (or orbital space), there appears a new state

$$\begin{aligned} A_{\mu j}(\hat{\boldsymbol{n}}, \theta) &= AR_{\mu j}(\hat{\boldsymbol{n}}, \theta) \\ R_{jl}(\hat{\boldsymbol{n}}, \theta) &= \cos\theta\delta_{jl} + (1 - \cos\theta)\hat{n}_j\hat{n}_l + \sin\theta\varepsilon_{jlk}\hat{n}_k. \end{aligned} \tag{6.15}$$

This class of states with phase χ and rotation specified by $\hat{\boldsymbol{n}}$ and θ is called the *BW state* (Balian and Werthamer 1963, see [B-7]). One of the characteristics of the BW state is that it is invariant under simultaneous rotations in spin and orbit spaces, which are in the subgroup H of G. The symmetry $SO_3 \times U(1)$ in Eq. (6.12) is broken in the BW state while the other symmetry SO_3 is left unbroken. In this sense, the BW state is the most symmetric state among possible ^{3}P states. Note however that the symmetry with respect to the relative angle between spin space and orbital space of the pair is broken. This manifests itself most clearly in the nuclear magnetic resonance mentioned later.

It can be readily shown from Eq. (6.15) that $\boldsymbol{\Delta}^* \times \boldsymbol{\Delta} = \boldsymbol{0}$, since $\boldsymbol{\Delta}(\boldsymbol{k})$ in Eq. (2.40) is proportional to $\boldsymbol{A}(\boldsymbol{k})$, and hence the BW state is unitary. Moreover the energy gap is isotropic since $\hat{\Delta}^\dagger\hat{\Delta} = \boldsymbol{\Delta}^* \cdot \boldsymbol{\Delta} = |\Delta|^2\hat{k}_j\hat{k}_j = 1$.

6.2.2 *ABM state*

Consider three mutually orthogonal unit vectors $\hat{\boldsymbol{l}}, \hat{\boldsymbol{m}}, \hat{\boldsymbol{n}}$ in momentum space such that $\hat{\boldsymbol{m}} \cdot \hat{\boldsymbol{n}} = 0$ and $\hat{\boldsymbol{l}} = \hat{\boldsymbol{m}} \times \hat{\boldsymbol{n}}$. Then the orbital state represented by $\hat{\boldsymbol{m}} + \mathrm{i}\hat{\boldsymbol{n}}$ is a state whose orbital angular momentum along $\hat{\boldsymbol{l}}$ is unity as shown in Fig. 6.4. Suppose both spin ↑↑ pair and spin ↓↓ pair take this orbital state as shown in Fig. 6.5. The components $A_{\mu j}$ when $\hat{\boldsymbol{m}} = \hat{\boldsymbol{x}}$ and $\hat{\boldsymbol{n}} = \hat{\boldsymbol{y}}$ are

$$(A_{\mu j}) = A\begin{pmatrix} 0 & 0 & 0 \\ 1 & \mathrm{i} & 0 \\ 0 & 0 & 0 \end{pmatrix}. \tag{6.16}$$

Although the vector A_μ in this form is directed along the y-axis in spin space, it may in general take any direction. Let us call this direction $\hat{\boldsymbol{d}}$. Thus one obtains a class of order parameters

$$A_{\mu j} = A\hat{d}_\mu(\hat{m}_j + \mathrm{i}\hat{n}_j) \tag{6.17}$$

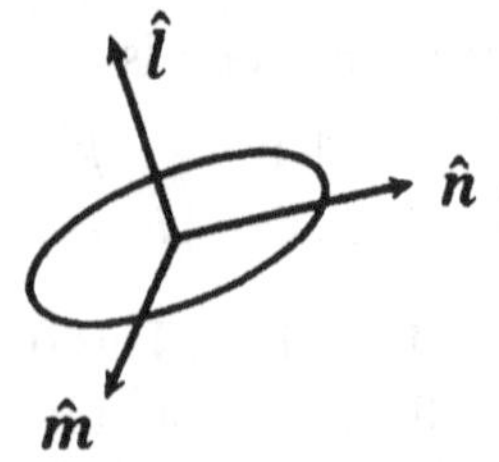

Fig. 6.4. The triad of three mutually orthogonal unit vectors.

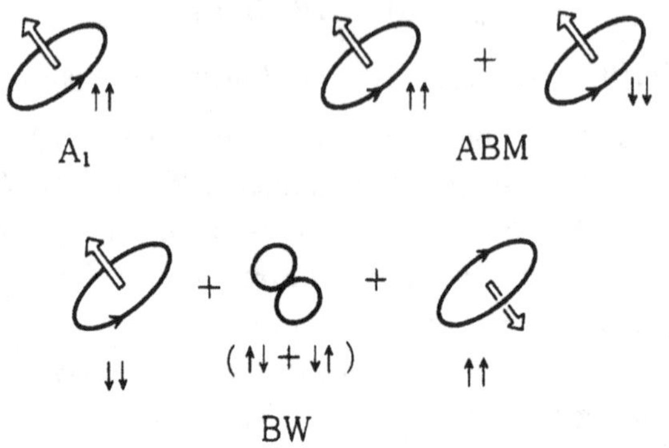

Fig. 6.5. The orbital state and the spin state in the A_1, ABM and BW states.

which is specified by a vector $\hat{\boldsymbol{d}}$ and a triad of unit vectors $(\hat{\boldsymbol{l}}, \hat{\boldsymbol{m}}, \hat{\boldsymbol{n}})$. This class is called the *ABM state* (Anderson–Brinkman–Morel 1973, see [B-7]). The orbital state transforms as $(\hat{\boldsymbol{m}} + i\hat{\boldsymbol{n}}) \to e^{i\chi}(\hat{\boldsymbol{m}} + i\hat{\boldsymbol{n}})$ under rotation of the triad along $\hat{\boldsymbol{l}}$ by an angle χ. Thus the state (6.17) is left invariant if this rotation is combined with a multiplicative gauge transformation $e^{-i\chi}$. In other words, gauge transformation is absorbed into a rotation of the triad. The state is also left invariant under rotations around the $\hat{\boldsymbol{d}}$-vector in spin space. Furthermore, Eq. (6.17) is left invariant under the simultaneous transformations $\hat{\boldsymbol{d}} \to -\hat{\boldsymbol{d}}$ and phase rotation by $\pi/2$. Therefore, the residual symmetries in the ABM state are products of the rotations around $\hat{\boldsymbol{d}}$, those around $\hat{\boldsymbol{l}}$ and the inversion of $\hat{\boldsymbol{d}}$, namely $U(1) \times U(1) \times Z_2$.

The ABM state is unitary too and it follows from Eq. (6.17) that the energy gap is proportional to $(\hat{\boldsymbol{m}} \cdot \hat{\boldsymbol{k}} - i\hat{\boldsymbol{n}} \cdot \hat{\boldsymbol{k}})(\hat{\boldsymbol{m}} \cdot \hat{\boldsymbol{k}} + i\hat{\boldsymbol{n}} \cdot \hat{\boldsymbol{k}}) = 1 - (\hat{\boldsymbol{l}} \cdot \hat{\boldsymbol{k}})^2$. Accordingly the gap vanishes at the poles along $\hat{\boldsymbol{l}}$ as shown in Fig. 6.6(a). This fact may be deduced directly from the symmetry of the ABM state as follows. If the operation $R(\hat{\boldsymbol{l}}, \pi)$ of the rotation along $\hat{\boldsymbol{l}}$ by π acts on Ψ, one has

$$R(\hat{\boldsymbol{l}}, \pi)\Psi(\hat{\boldsymbol{k}}) = \Psi(R(\hat{\boldsymbol{l}}, \pi)\hat{\boldsymbol{k}}).$$

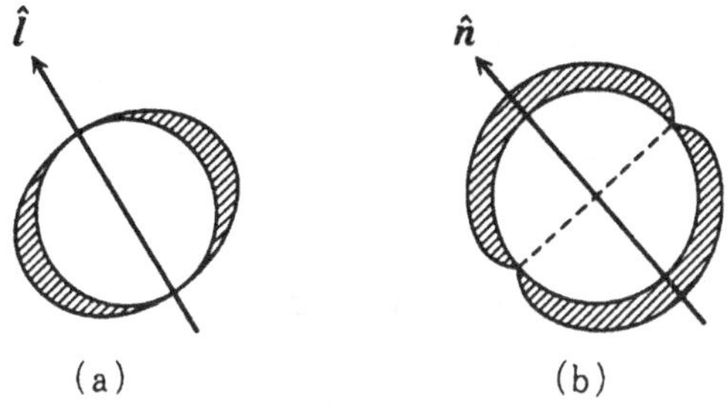

Fig. 6.6. The energy gap in (a) the ABM state and (b) the planar state.

This transformation is equivalent to multiplication of the phase $e^{i\pi}$ of $U(1)$ as mentioned above. If $\hat{\boldsymbol{k}}$ is equal to $\hat{\boldsymbol{k}}_0 \parallel \hat{\boldsymbol{l}}$, one finds $\Psi(\hat{\boldsymbol{k}}_0) = 0$ since $R(\hat{\boldsymbol{l}}, \pi)\hat{\boldsymbol{k}}_0 = \hat{\boldsymbol{k}}_0$ implies $\Psi(\hat{\boldsymbol{k}}_0) = -\Psi(\hat{\boldsymbol{k}}_0)$. The nodes in the energy gap may be obtained from group theoretical consideration as can be guessed from this example. This kind of reasoning is employed when the gap in a heavy fermion system is considered in Chapter 7.

6.2.3 A_1 state

A state with spin $\uparrow\uparrow$ pairs only, called the A_1 state, is obtained if Eq. (6.16) is replaced by

$$(A_{\mu j}) = A \begin{pmatrix} -i & 1 & 0 \\ 1 & i & 0 \\ 0 & 0 & 0 \end{pmatrix}. \tag{6.18}$$

If rotations in spin space and $\boldsymbol{k}$-space are applied separately on the above state, one obtains an order parameter

$$(A_{\mu j}) = A(\hat{d}_\mu + i\hat{e}_\mu)(\hat{m}_j + i\hat{n}_j) \tag{6.19}$$

with $\hat{\boldsymbol{d}} \perp \hat{\boldsymbol{e}}$. This time, so far as rotations in spin space are concerned, the state is invariant only under rotations round the vector $\hat{\boldsymbol{d}} \times \hat{\boldsymbol{e}}$. The A_1 state is of course non-unitary. This state, in which spins in a particular direction form pairs while oppositely directed spins remain in a normal state without the energy gap, describes the A_1 phase which is realised in a magnetic field.

6.2.4 Polar state and planar state

Let $\hat{\boldsymbol{d}}$ and $\hat{\boldsymbol{n}}$ be unit vectors in spin space and $\boldsymbol{k}$-space respectively. Then the state whose order parameter is

$$A_{\mu j} = A\hat{d}_\mu \hat{n}_j \tag{6.20}$$

is called the *polar state* since the orbit extends along the vector $\hat{\boldsymbol{n}}$. Because the energy gap is proportional to $(\hat{\boldsymbol{n}}\cdot\hat{\boldsymbol{k}})^2$, it vanishes on the equator perpendicular to $\hat{\boldsymbol{n}}$ on the Fermi sphere as shown in Fig. 6.6(b).

If the order parameter is given by

$$A_{\mu j} = A(R_{\mu j} - R_{\mu k}\hat{l}_k\hat{l}_j) \tag{6.21}$$

in terms of $R_{ij}(\hat{\boldsymbol{n}}, \theta)$ in Eq. (6.15), the state is called the *planar state* since the orbit extends perpendicular to $\hat{\boldsymbol{l}}$.

6.2.5 Other superfluid states

Other superfluid states with different symmetry are possible. Their group theoretical classification is given in detail in [B-7]. In connection with the superfluidity of ^{3}He we will mainly consider physical properties of the states described in Sections 6.2.1–6.2.3. It should be remarked that the $^3P_2$ pairing is considered to be responsible for the superfluidity of neutron liquid in a neutron star since the nuclear force contains a strong tensor force as previously mentioned. In other words it is a superfluid whose spin and orbits are fixed in the $J = 2$ state. It might be thought in this case that each pair carries an angular momentum of $2\hbar$ and hence the system carries an intrinsic angular momentum of $2\hbar \cdot n_s/2$ per unit volume. It must be understood, however, that particles form a pair outside the Fermi sphere while holes form a pair inside the Fermi sphere and there is therefore net contribution to the intrinsic angular momentum due to deviation from particle–hole symmetry. Therefore the magnitude of the intrinsic angular momentum is reduced by a factor Δ/ε_F. This argument applies equally to the pair orbital angular momentum of the ABM state. For the angular momentum of the whole system, however, one must include the contribution from superfluid flow near the container wall, and no simple conclusion can then be drawn from the present argument; see Section 6.6.

6.3 Physical properties of the ^{3}P superfluid states

The expressions for the free energy, the gap equation and so on within the weak coupling theory are of the same structure in a superfluid state regardless of the pairing, as shown in Chapter 2. Therefore the results for liquid ^{3}He with a ^{3}P pairing will be obtained if the theory of ^{1}S superconductivity in Chapter 3 is generalised *mutatis mutandis*.

6.3.1 Energy gap

Let us consider the energy gap touched upon in the previous section in more detail. If the interaction (6.10) is assumed, Eq. (2.58) becomes

$$\Delta_{\mu i} = 3g_1 \sum_{k'} \mathrm{tr}\left\{\frac{1}{2} i\sigma_2\sigma_\mu \hat{\Delta}(\hat{k}')\hat{k}_i{}' \frac{\tanh(\beta\hat{\varepsilon}_{k'}/2)}{2\hat{\varepsilon}_{k'}}\right\}. \tag{6.22}$$

In a unitary state this simplifies to

$$\Delta_{\mu i} = 3g_1 \sum_n \Delta_{\mu n} \sum_{k'} \hat{k}_i{}'\hat{k}_n{}' \frac{\tanh(\beta\varepsilon_{k'}/2)}{2\varepsilon_{k'}}. \tag{6.23}$$

We must keep in mind that the gap in the quasi-particle excitation energy ε_k is in general a function of $\hat{\boldsymbol{k}}$. In a unitary state, this gap is given by $|\Delta_k|^2 = \frac{1}{2}\mathrm{tr}(\Delta_k^\dagger \Delta_k)$. The critical temperature T_c is given, in any ^{3}P state, by

$$k_B T_c = 1.13\,\omega_c e^{-1/N(0)g_1} \tag{6.24}$$

just as in the BCS theory, as was mentioned at the end of Chapter 2.

1. **BW state** The gap in the BW state is given by $\Delta_{\mu j} = \Delta R_{\mu j}(\hat{\boldsymbol{n}}, \theta)$, see Eq. (6.15), and the energy gap is isotropic, $|\Delta_k|^2 = |\Delta|^2$. Therefore Eq. (6.23) is identical to that for the ^{1}S pairing and the temperature dependence of Δ agrees with the BCS result. It should be noted that the only case in which an isotropic gap is produced in $l \neq 0$ pairings is the BW state in the ^{3}P pairing.
2. **ABM state** In Eq. (6.17) one may take $\hat{\boldsymbol{m}} = \hat{\boldsymbol{x}}$ and $\hat{\boldsymbol{n}} = \hat{\boldsymbol{y}}$, without loss of generality. Then the gap equation (6.23) becomes

$$1 = N(0)g_1 \frac{3}{4} \int_{-\omega_c}^{\omega_c} d\xi \int_{-1}^{1} d\mu (1-\mu^2) \frac{\tanh(\beta\varepsilon/2)}{2\varepsilon}, \tag{6.25}$$

where $\varepsilon = \sqrt{\xi^2 + \Delta_A^2(1-\mu^2)}$ and $\mu = \hat{\boldsymbol{k}} \cdot \hat{\boldsymbol{z}}$. The above integral is evaluated as $\ln(2\omega_c/\Delta) + 5/6 - \ln 2$ at $T = 0$ and Δ_A is given by

$$\Delta_A = \Delta_0 e^{0.14} \tag{6.26}$$

where Δ_0 is the gap of the BW state at $T = 0$. The condensation energy is proportional to the angular average of $|\Delta_A|^2(1-\mu^2)$ over the Fermi surface, which becomes $(2/3)e^{0.28}\Delta_0^2 = 0.88\Delta_0^2$. Accordingly the BW state has lower energy in the weak coupling theory. In fact, the B phase with an isotropic gap, identified with the BW state, appears at low temperatures. Weak coupling theory, however, cannot explain why the A phase, identified with the ABM state, appears just below T_c, since it predicts that the BW state is stable even near T_c as will be shown in Section 6.5. It should be noted that the pair $\uparrow\uparrow$ is independent of the pair $\downarrow\downarrow$ in the ABM state in the weak coupling limit and there is no reason why they should take the same orbital state. The reason for this must be sought in the strong coupling effect.

Finally the gap in the planar state is the same as that in the ABM state and the gap in the polar state is given by $|\Delta_k|^2 = |\Delta_{\rm pl}|^2(\hat{\boldsymbol{k}}\cdot\hat{\boldsymbol{l}})^2$. The amplitude at $T = 0$ is $|\Delta_{\rm pl}| = {\rm e}^{1/3}\Delta_0$. Therefore $\overline{|\Delta_k|^2} = ({\rm e}^{2/3}/3)\Delta_0^2$ in the polar state and its condensation energy is the lowest.

6.3.2 *Specific heat*

The energy gap profile is reflected in the temperature dependence of the specific heat C. Once the quasi-particle energy ε_k is known, the specific heat C is obtained from the entropy (2.57). Let us make a qualitative consideration of the temperature dependence of C at $T \ll T_{\rm c}$ leaving aside the case with $T \sim T_{\rm c}$ to Section 6.5. First of all, the gap in the BW state is isotropic and C is identical to the BCS result derived in Section 3.2, and hence C decreases exponentially at low temperatures. In contrast, the gap in the ABM state of the planar state varies as $\Delta_{\rm A}\sqrt{1-\mu^2}$ on the Fermi surface with $\mu = \hat{\boldsymbol{k}}\cdot\hat{\boldsymbol{l}}$ and vanishes near the poles $\mu = \pm 1$. Therefore only the quasi-particles within the angular domain $\sqrt{1-\mu^2} < \varepsilon/\Delta_{\rm A}$ are excited by energies ε ($\ll \Delta_{\rm A}$). The number of quasi-particles involved is of the order of $N(0)\varepsilon\cdot(\varepsilon/\Delta_{\rm A})^2$. Then from the above consideration one estimates that

$$C \propto N(0){k_{\rm B}}^4T^3/\Delta_{\rm A}^2. \tag{6.27}$$

Calculation shows that the coefficient of the above equation is $7\pi^2/5$. Similar calculations for the polar state, for which $\Delta = 0$ on the equator, yield the density of states $N(0)\varepsilon(\varepsilon/\Delta_{\rm pl})$ and specific heat proportional to T^2. The existence of a specific heat proportional to a power of the temperature in a certain class of states gives an important clue to understanding superconductivity in heavy electron systems as discussed in Chapter 7.

6.3.3 *Superfluid component*

The expression (3.39) for the number density contributing to superfluidity is general, in so far as the quasi-particles can be regarded as elementary excitations, and may be applicable to any ^{3}P superfluid state. It should be noted, however, that the magnitude of the superfluid component depends

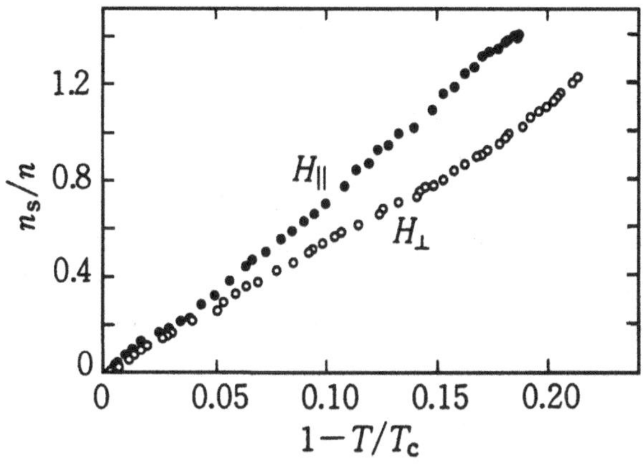

Fig. 6.7. The temperature dependence of the superfluid components with the flow parallel and perpendicular to an applied magnetic field. (From [G-18].)

on the direction $\boldsymbol{q}$ relative to the centre of mass motion in general since the energy gap $|\Delta_k|^2$ is anisotropic. In other words, n_s is a tensor with components

$$n_{sij} = n\delta_{ij} - \frac{1}{m}\sum_k \frac{\partial f(\varepsilon_k)}{\partial \varepsilon_k}\hat{k}_j\hat{k}_j. \tag{6.28}$$

Since the BW state has an isotropic gap, n_s in this case is identical to that in the BCS theory. In the ABM state, on the other hand, the gap depends on the relation between $\hat{\boldsymbol{l}} = \hat{\boldsymbol{m}} \times \hat{\boldsymbol{n}}$ and $\boldsymbol{q}$. One carries out the calculation near T_c in a similar way to that for Eq. (3.41) to obtain

$$\begin{aligned} \frac{n_{s\parallel}}{n} &= \frac{7\zeta(3)}{10\pi^2 {T_c}^2}\Delta_A^2(T) \\ n_{s\perp} &= 2n_{s\parallel}. \end{aligned} \tag{6.29}$$

In other words, the normal fluid component when the flow is parallel to $\hat{\boldsymbol{l}}$ is larger than that when the flow is normal to $\hat{\boldsymbol{l}}$. This result is not unexpected since the gap vanishes along $\pm\hat{\boldsymbol{l}}$ on the Fermi surface and hence there are large numbers of thermally excited quasi-particles around these points. The direction of the vector $\hat{\boldsymbol{l}}$ is controllable as it is perpendicular to the wall and also perpendicular to an applied magnetic field. Figure 6.7 shows experimental results on superfluid density in the A phase, which is one of the proofs identifying the A phase with the ABM state.

6.3.4 Magnetisation

The susceptibility associated with the nuclear spin of ^{3}He well demonstrates the characteristics of the spin-triplet pairing. Suppose the direction of an external magnetic field is taken as the quantisation axis. Among the triplet of spin states, $\uparrow\uparrow$, $\downarrow\downarrow$ and $\uparrow\downarrow + \downarrow\uparrow$, the third one is made of oppositely directed spins and hence we expect the same result for this component as the ^{1}S case. On the other hand, it seems that one only needs to make the simple replacements $\xi_k \to \xi_k \pm \gamma B/2$ for the components $\uparrow\uparrow$ and $\downarrow\downarrow$. In fact, this can be easily verified by examining Eqs. (2.50) and (2.51) with $\hat{\xi}_k = \xi_k \hat{1} + \sigma_3 \gamma B/2$. (If one puts $\Delta \propto i\sigma_3\sigma_2$, $\gamma B/2$ does not appear in Eq. (2.50) as in the spin-singlet case.) Since there are only $\uparrow\uparrow$ and $\downarrow\downarrow$ pairs in the ABM state with $\hat{\boldsymbol{d}} \perp \boldsymbol{B}$, pairing with these components is possible although the number of up spins differs from that of down spins by $N(0)\gamma B/2$ just as in the normal state. Accordingly one finds $\chi = \chi_{\mathrm{n}} = \gamma^2 N(0)/2$ in this case. On the other hand, one has the same result as the ^{1}S case if $\hat{\boldsymbol{d}}$ is parallel to the magnetic field $\boldsymbol{B}$. The $\hat{\boldsymbol{d}}$-vector satisfies $\hat{\boldsymbol{d}} \perp \boldsymbol{B}$ in a magnetic field if there is no other mechanism that fixes $\hat{\boldsymbol{d}}$ and hence one should obtain $\chi = \chi_{\mathrm{n}}$. In fact it is observed that χ does not change at the transition to the A phase.

Next we consider the BW state. It follows from the above consideration that the part $\hat{k}_x^2 + \hat{k}_y^2$ yields χ_{n} while the part $\hat{k}_z^2$ yields the same χ_{s} as that for the ^{1}S pairing given by Eq. (3.43). Therefore the susceptibility in the BW state is given by

$$\chi_{\mathrm{BW}} = \frac{1}{3}(2\chi_{\mathrm{n}} + \chi_{\mathrm{s}}). \tag{6.30}$$

We expect the above susceptibility to approach $2\chi_{\mathrm{n}}/3$ as the temperature is lowered since χ_{s} vanishes at $T = 0$. Figure 6.8 shows that χ in the B phase decreases with decreasing temperature, approaching a finite limit. This limiting value deviates from 2/3, however, and for quantitative agreement, the Fermi liquid effects should be considered, as is the case also for n_{s}.

Liquid ^{3}He undergoes a transition from the normal state to the A phase through the A_1 phase in a magnetic field as shown in Fig. 6.1(a). The number density of particles with spin parallel to the magnetic field is larger than that with the opposite spin by $N(0)\gamma B$ in the normal state. Therefore the densities of states at the respective Fermi surfaces differ as $N_{\uparrow(\downarrow)}(0) = N(0) + (-) m^2\gamma B/2\pi^2 k_{\mathrm{F}}$ and, as a consequence, the A_1 phase with $\uparrow\uparrow$ pair has a higher T_{c}. A rough estimate of the difference in T_{c} is obtained from Eq. (6.24) as

$$\frac{T_{\mathrm{c}\uparrow} - T_{\mathrm{c}\downarrow}}{T_{\mathrm{c}}} \sim \frac{1}{N(0)g_1} \cdot \frac{\gamma B}{\varepsilon_{\mathrm{F}}}, \tag{6.31}$$

where the change in the interaction has been ignored.

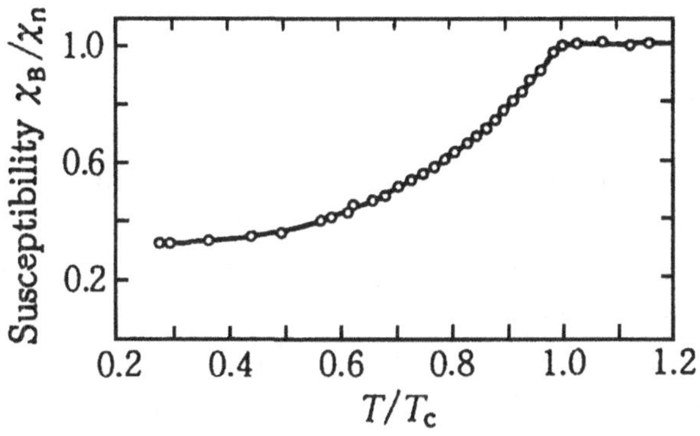

Fig. 6.8. The temperature dependence of the susceptibility at 20 bar in the B phase. (From [G-19].)

6.3.5 Fermi liquid effects

The pairing itself does not change the particle and spin densities if there is particle–hole symmetry above and below the Fermi surface. Accordingly the Fermi liquid effects do not appear in the gap equation for example. However the expressions for, say, the normal component and the susceptibility must be modified due to the Fermi liquid effects since the way the medium polarises changes substantially under the superfluid transition. This is most easily understood in the susceptibility. In the superfluid state one needs to put χ_s^0 in Eq. (6.6) in place of χ_n^0 which corresponds to $\partial f_{p\alpha}^0/\partial\xi_{p\alpha}$. Here χ^0 is the susceptibility without the Fermi liquid effects. Then one obtains

$$\frac{\chi_s}{\chi_n} = \frac{1 + F_0^{(a)}}{1 + (\chi_s^0/\chi_n^0)F_0^{(a)}} \cdot \frac{\chi_s^0}{\chi_n^0}. \tag{6.32}$$

Therefore one finds $\chi_s = \chi_n$ in the ABM state while

$$\frac{\chi_s}{\chi_n} = \frac{2}{3}\frac{1 + F_0^{(a)}}{1 + 2F_0^{(a)}/3}$$

as $T \to 0$ in the BW state. Figure 6.8 clearly shows this effect.

The correction to the normal component can be obtained in a similar way, resulting in the expression

$$n_n = \frac{n_n^0}{1 + (mn_n^0/m^*n)(F_1^{(s)}/3)}. \tag{6.33}$$

6.4 Spin dynamics and nuclear magnetic resonance

One of the remarkable phenomena in superfluid ^{3}He is nuclear spin dynamics. As already noted, the dominant interatomic force in ^{3}He is invariant under separate rotations in spin space and real (momentum) space. However superfluid ordering breaks the relative symmetry. This spin–orbit symmetry breaking manifests itself most clearly in the nuclear spin dynamics since there is a very weak spin–orbit coupling interaction between the nuclear spins, called the magnetic dipole interaction H_{D}, or the d–d interaction. The interaction H_{D} between nuclear magnetic moments $\frac{1}{2}\gamma\boldsymbol{\sigma}$ is written as

$$H_{\mathrm{D}} = -\frac{\pi}{3}\left(\frac{\gamma}{2}\right)^2 \sum_{k,k',q} \left(\frac{3q_iq_j}{\boldsymbol{q}^2} - \delta_{ij}\right) \mathrm{tr}\,(\hat{a}_k^\dagger \sigma_i \hat{a}_{k+q})\mathrm{tr}\,(\hat{a}_{k'}^\dagger \sigma_j \hat{a}_{k'-q}), \qquad (6.34)$$

where tr stands for the trace over spin indices. The statistical average in a specific superfluid state $E_{\mathrm{D}} = \langle H_{\mathrm{D}} \rangle$ may be obtained in the mean field approximation. Thus, if the order parameter is written as $\Psi(\hat{\boldsymbol{k}})$, one obtains

$$E_{\mathrm{D}} = -\frac{\pi}{3}\left(\frac{\gamma}{2}\right)^2 \sum_{k,k'} \left\{\frac{3(\hat{\boldsymbol{k}} - \hat{\boldsymbol{k}}')_i(\hat{\boldsymbol{k}} - \hat{\boldsymbol{k}}')_j}{|\hat{\boldsymbol{k}} - \hat{\boldsymbol{k}}'|^2} - \delta_{ij}\right\} \mathrm{tr}\,(\sigma_i \Psi^\dagger(\boldsymbol{k}) \sigma_j^{\mathrm{T}} \Psi(\boldsymbol{k}')),$$

where the approximation $|\boldsymbol{k}| \sim k_{\mathrm{F}}$ is employed as usual. If the representation $\Psi = A_\mu \mathrm{i}\sigma_\mu \sigma_2$ is used, one obtains $\mathrm{tr}(\cdots) = -2(A_i A_j^* + A_i A_j^* - \delta_{ij} A_\lambda A_\lambda^*)$. Then substitute Eq. (6.11) into the above equation and integrate the resulting expression with respect to $\hat{\boldsymbol{k}}, \hat{\boldsymbol{k}}'$ and ξ, ξ'. These integrations are easily carried out if one takes three contractions with respect to the four components of $\hat{\boldsymbol{k}}$ and $\hat{\boldsymbol{k}}'$ since only scalar quantities remain after integrations with respect to angles. Finally one finds

$$E_{\mathrm{D}} = \frac{\pi}{20}\gamma^2 \frac{1}{g_1^2}\left(\Delta_{ii}\Delta_{jj}^* + \Delta_{ij}\Delta_{ji}^* - \frac{2}{3}\Delta_{ij}\Delta_{ij}^*\right), \qquad (6.35)$$

where the notation $\Delta_{ij} = N(0) g_1 \int \mathrm{d}\xi A_{ij}(\xi)$ has been introduced. The magnitude of the above equation is estimated as $E_{\mathrm{D}} \sim n(\Delta_0/\varepsilon_{\mathrm{F}})^2 (n\gamma^2)$ since $1/N(0)g_1$ is of the order of unity.

The energy E_{D} of course depends on the state. For the ABM state, if Eq. (6.17) is substituted into Eq. (6.35), one obtains

$$E_{\mathrm{D}} = \frac{\pi}{10}\frac{\Delta_{\mathrm{A}}^2 \gamma^2}{g_1^2}\left[\frac{1}{3} - (\hat{\boldsymbol{l}} \cdot \hat{\boldsymbol{d}})^2\right]. \qquad (6.36)$$

While for the BW state, if Eq. (6.15) is substituted into Eq. (6.35) one obtains

$$E_{\rm D} = \frac{2\pi}{15}\frac{\Delta_{\rm B}{}^2}{g_1{}^2}\gamma^2\left[2\left(\cos\theta + \frac{1}{4}\right)^2 + \frac{5}{8}\right]. \tag{6.37}$$

This expression is independent of $\hat{\boldsymbol{n}}$ since $\hat{\boldsymbol{n}}$ is the only vector in Eq. (6.15).

In contrast with the condensation free energy which is independent of $\hat{\boldsymbol{d}}\cdot\hat{\boldsymbol{l}}$ or θ, the energy $E_{\rm D}$ depends on these parameters representing the 'breaking of the symmetry' although the magnitude of the latter is far smaller than that of the former. This dependence is observed directly by nuclear magnetic resonance (NMR).

The energy $E_{\rm D}$ is known to vanish by symmetry in a liquid or in a crystal with a cubic symmetry even with a ferromagnetic order. It remains finite in a superfluid state of ^{3}He, though it is a fluid, since the relative orientation of the position vector and the triplet spin vector is fixed, that is to say, the spin–orbit symmetry is broken.

6.4.1 Spin dynamics and NMR

The spin motion in a superfluid state is a form of collective mode. A small amplitude oscillation with a finite wave number is just a spin wave. Here, uniform motion with wave number 0, that is, the motion of the total spin angular momentum $\boldsymbol{S}$ is considered, since nuclear magnetic resonance is under consideration. One should recall that $\boldsymbol{S}$ is a classical quantity and the generator of infinitesimal rotations in the spin space of the system. Accordingly the canonical conjugate quantity to $\boldsymbol{S}$ is a vector $\boldsymbol{\theta}(=\hat{\boldsymbol{n}}\theta)$ representing a rotation in spin space and the equations of motion in an external field $\boldsymbol{H}$ are

$$\begin{aligned} \frac{\mathrm{d}\boldsymbol{S}}{\mathrm{d}t} &= \gamma\boldsymbol{S}\times\boldsymbol{H} - \frac{\partial E_{\rm D}}{\partial\boldsymbol{\theta}} \\ \frac{\mathrm{d}\boldsymbol{\theta}}{\mathrm{d}t} &= -\gamma(\boldsymbol{H} - \chi^{-1}\gamma\boldsymbol{S}). \end{aligned} \tag{6.38}$$

The second equation is just Larmor's theorem, which states that the effective magnetic field in a rotating system with angular velocity $\boldsymbol{\omega} = \mathrm{d}\boldsymbol{\theta}/\mathrm{d}t$ is $\boldsymbol{H}+\gamma^{-1}\boldsymbol{\omega}$. It is assumed that the spin motion is slow enough for susceptibility and $E_{\rm D}$ in an equilibrium state to be used. If $\boldsymbol{S}$ is eliminated from Eqs. (6.38), the equation of motion obtained is

$$\frac{\mathrm{d}^2\boldsymbol{\theta}}{\mathrm{d}t^2} - \gamma\frac{\mathrm{d}\boldsymbol{\theta}}{\mathrm{d}t}\times\boldsymbol{H} + \gamma^2\chi^{-1}\frac{\partial E_{\rm D}}{\partial\boldsymbol{\theta}} = -\gamma\frac{\mathrm{d}\boldsymbol{H}}{\mathrm{d}t}. \tag{6.39}$$

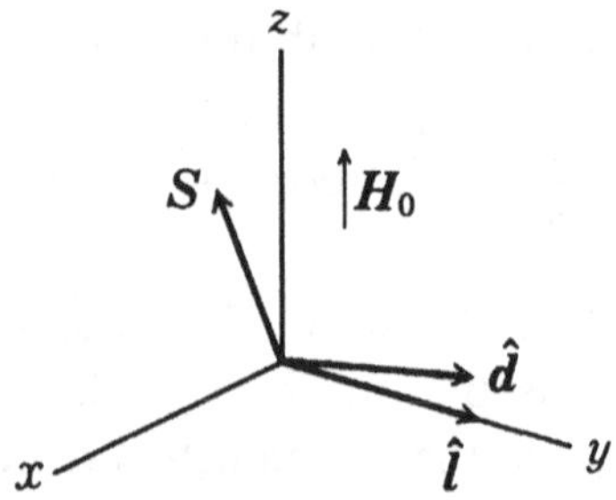

Fig. 6.9. Spin motion in the ABM state.

ABM state

Explicit calculation is easily carried out for the ABM state identified with the A phase. The $\hat{\boldsymbol{d}}$ vector is perpendicular to a uniform static magnetic field $\boldsymbol{H}_0$ as remarked in the analysis of the susceptibility. Furthermore, the most stable state is realised when $\hat{\boldsymbol{d}} \perp \boldsymbol{H}_0$ and $\hat{\boldsymbol{d}} \parallel \hat{\boldsymbol{l}}$ since E_D in Eq. (6.36) is minimised when $\hat{\boldsymbol{l}}$ is parallel to $\hat{\boldsymbol{d}}$. Let us consider small oscillations about this state. It may be assumed that the orbital part of the pair, namely the vector $\hat{\boldsymbol{l}}$, is kept fixed.

The coordinate axes are taken as $\boldsymbol{H}_0 \parallel \hat{z}$ and $\hat{\boldsymbol{l}} \parallel \hat{y}$, see Fig. 6.9. Equation (6.39) separates into transverse oscillations, in which θ_x and θ_y change, and a longitudinal oscillation, in which θ_z changes, since for a small oscillation $\boldsymbol{\theta}$ is given by a summation of respective rotation angles around each axis and hence $(\hat{\boldsymbol{l}} \cdot \hat{\boldsymbol{d}}) = 1 - \frac{1}{2}(\theta_x^2 + \theta_y^2)$. The resonance frequencies are

$$\begin{aligned} \omega^2 &= \omega_0{}^2 + \Omega_A{}^2 \\ \omega^2 &= \Omega_A{}^2, \end{aligned} \tag{6.40}$$

where $\omega_0 = \gamma H_0$ and

$$\Omega_A{}^2 \equiv \frac{\pi}{5} \frac{\gamma^4 \Delta_A{}^2}{\chi_A g_1{}^2} \tag{6.41}$$

is the shift in the transverse, i.e., ordinary NMR resonance frequency, see Fig. 6.10. Longitudinal oscillation is oscillation in the magnitude of $\boldsymbol{S}$, characteristic of the superfluid state. This mode is called ringing when the amplitude is finite and can be regarded as a form of a.c. Josephson effect in which the weak d–d interaction couples the ↑↑ and ↓↓ superfluids.

BW state

The variables which represent the breaking of the spin–orbit symmetry in the BW state corresponding to the B phase are $\hat{\boldsymbol{n}}$ and θ in Eq. (6.15). The

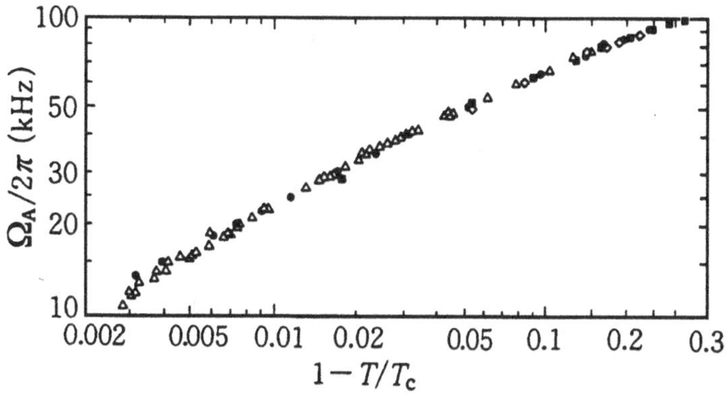

Fig. 6.10. The temperature dependence of the shift Ω_A in the NMR resonance frequency in the A phase under melting pressure. The data obtained from transverse resonance are denoted by △ while those from longitudinal resonance are • and ■. (D.M. Lee and R.C. Richardson, from [B-1].)

susceptibility χ is isotropic, however, and the direction $\hat{\boldsymbol{n}}$ cannot be fixed from magnetic energy considerations. Moreover the energy E_D of Eq. (6.37) does not fix $\hat{\boldsymbol{n}}$ either. It can be shown, however, that a term proportional to H^2 in E_D is minimised when $\hat{\boldsymbol{n}} \parallel \boldsymbol{H}$. Although this E_D contribution is even smaller than that in Eq. (6.37), it makes $\hat{\boldsymbol{n}}$ parallel or anti-parallel to $\boldsymbol{H}$ in the uniform equilibrium state. The angle θ in Eq. (6.37) is, then, the rotation angle θ_z around the axis $\hat{\boldsymbol{z}}$ ($\parallel \boldsymbol{H}_0$). Therefore there is no torque due to E_D acting on the transverse oscillations and hence the resonance frequency does not shift from $\omega_0 = \gamma H_0$. Equation (6.37) is minimised at $\theta_0 = \cos^{-1}(-1/4) = 104°$ and $2\pi - \theta_0$. The frequency of a small oscillation of θ_z around the equilibrium angle, namely the frequency of the longitudinal oscillation is

$$\Omega_B{}^2 = \frac{\pi}{2}\frac{\gamma^4 \Delta_B{}^2}{\chi_B g_1{}^2}. \tag{6.42}$$

The ratio of the above frequency to that of the A phase is

$$(\Omega_B/\Omega_A)^2 = (5/2)(\Delta_B/\Delta_A)^2(\chi_A/\chi_B), \tag{6.43}$$

which has been verified experimentally near the A–B transition point.

The frequencies Ω_A and Ω_B are of the order of a few kHz. Although E_D is fairly small, the shift in the resonance frequency is easily observable since χ, corresponding to the inertial mass in the above equation of motion, is small. (A shift due to the same mechanism has been observed in the u2d2 phase in solid ^{3}He with nuclear spin ordering.) At finite temperatures only the resonance at $\omega = \sqrt{\omega_0^2 + \Omega_A^2}$ is observed even in the presence of thermal excitations. This is because unpaired particles also receive torque from E_D.

6.5 Ginzburg–Landau theory

It has been shown in Chapter 5 that the Ginzburg–Landau theory is effective for practical analysis of many phenomena in ordinary superconductivity. This approach is indispensable in the more complicated ^{3}P superfluid. In the Ginzburg–Landau theory, free energy $F_s - F_n$ is expanded in terms of the energy gap or the order parameter assuming $T \sim T_c$. Terms up to the fourth order in the order parameter and terms up to the second order in the spatial derivative of the order parameter are kept in the above expansion, assuming that the scale of the spatial variation is larger than the coherence length ξ_0. One may readily write down the form of the free energy since it must be invariant (a scalar) under transformations associated with symmetry breaking, namely spin space rotation, orbital space rotation and gauge transformation. First of all, the second order term in the order parameter must be a scalar in spin and orbital spaces separately and invariant under a global phase change and hence it is singled out to be $A^*_{\mu i}A_{\mu i}$. The same reasoning yields the free energy density in the form

$$\begin{aligned} f_s - f_n = \frac{1}{3}N(0)\Delta_e^2(T)\Big\{&-\alpha A^*_{\mu i}A_{\mu i} + \frac{1}{5}[\beta_1|A_{\mu i}A_{\mu i}|^2 + \beta_2(A^*_{\mu i}A_{\mu i})^2 \\ &+\beta_3 A^*_{\mu i}A_{\mu j}A^*_{\nu i}A_{\nu j} + \beta_4 A^*_{\mu i}A_{\mu j}A^*_{\nu j}A_{\nu i} + \beta_5 A^*_{\mu i}A^*_{\mu j}A_{\nu i}A_{\nu j}] \\ &+K_1\partial_i A^*_{\mu i}\partial_j A_{\mu j} + K_2\partial_i A^*_{\mu j}\partial_i A_{\mu j} + K_3\partial_i A^*_{\mu j}\partial_j A_{\mu i}\Big\}, \end{aligned} \tag{6.44}$$

where $\Delta_e^2(T) = [8\pi^2/7\zeta(3)](1 - T/T_c)T_c^2$. The sum of terms quadratic in $\partial_i A_{\mu j}$ is called the gradient energy. The coefficients α, β_i $(1 \leq i \leq 5)$ and K_i $(1 \leq i \leq 3)$ are constants determined solely by the properties of the normal state and independent of the order parameter. Moreover, these constants are obtained in the weak coupling theory, in which the pairing interaction is assumed to be a constant g_1 independent of the order parameter, as was done for superconductivity in Section 3.2. The gap equation (6.22) must be expanded in the power of $\hat{\Delta}_k^\dagger\hat{\Delta}_k$ using Eq. (3.10) to find α and β. For example, the third order term in the equation for $\Delta_{\mu i}$ is

$$\begin{aligned} 3g_1 N(0)(-\beta_c{}^3)\Delta_{\nu l}\Delta^*_{\lambda m}\Delta_{\eta n}\mathrm{tr}\left(\frac{1}{2}s_\mu{}^\dagger s_\nu s_\lambda{}^\dagger s_\eta\right)\int\frac{\mathrm{d}\Omega'}{4\pi}\hat{k}_i'\hat{k}_l'\hat{k}_m'\hat{k}_n' \\ \times\sum_n\int \mathrm{d}\xi'\frac{1}{[(2n+1)^2\pi^2 + \beta_c{}^2\xi'^2]^2}, \end{aligned}$$

where $s_\mu = \mathrm{i}\sigma_\mu\sigma_2$. The above equation is easily evaluated to yield β. Similarly the coefficients K are obtained if the gap equation with centre of mass momentum $\boldsymbol{q}$ of a pair is expanded in terms of $v_F\hat{\boldsymbol{k}}\cdot\boldsymbol{q}$. In summary, the coefficients in Eq. (6.44) are obtained within the weak coupling theory as

$$\begin{aligned}&\alpha = 1,\\ &\beta_1 = -1/2, \quad \beta_2 = \beta_3 = \beta_4 = 1, \quad \beta_5 = -1,\\ &K_1 = K_2 = K_3 = \xi^2/5,\end{aligned} \tag{6.45}$$

where $\xi^2 = v_F^2/6\Delta_e^2(T)$ is the coherence length defined by Eq. (5.3). If we use the order parameters of the BW (Eq. (6.15)) or the ABM state (Eq. (6.16)), we get the free energy of the respective state measured from the normal value as

$$\begin{aligned}\Delta f_{\mathrm{BW}}/N(0)\Delta_e^2 &= -\frac{5}{4}[3(\beta_1+\beta_2)+\beta_3+\beta_4+\beta_5]^{-1}\\ \Delta f_{\mathrm{ABM}}/N(0)\Delta_e^2 &= -\frac{1}{2}(\beta_2+\beta_4+\beta_5)^{-1}.\end{aligned} \tag{6.46}$$

If the weak coupling values are used, these numbers turn out to be $-3/5$ and $-1/2$, so that the BW state has lower free energy and should be stable.

6.5.1 Momentum density associated with superfluidity

One assigns a virtual 'charge' $2m$ to the ^{3}He pair and obtains the current density by taking the derivative of the free energy in the presence of a vector potential $\boldsymbol{A}$ with respect to $\boldsymbol{A}$ to find the momentum (mass flow) density associated with a superfluid flow. That is, one makes the replacement $\nabla \to \nabla - \mathrm{i}2m\boldsymbol{A}$ in Eq. (6.44) and then takes the derivative with respect to the ith component of $\boldsymbol{A}$ to find

$$j_{si} = 2m\frac{1}{3}N(0)\Delta_e^2(T)\,\mathrm{Im}\{K_1A_{\mu i}^*\partial_jA_{\mu j} + K_2A_{\mu j}^*\partial_iA_{\mu j} + K_3A_{\mu j}^*\partial_jA_{\mu i}\}. \tag{6.47}$$

6.5.2 Magnetic free energy

Next an additional term in the presence of an external magnetic field $\boldsymbol{H}$ is obtained. The quadratic invariant with respect to $\boldsymbol{H}$ and Δ is

$$f_H = \alpha_B H_\mu A_{\mu i}{}^* H_\nu A_{\nu i}, \tag{6.48}$$

where α_B is a constant. This energy must be a term proportional to $\Delta^\dagger\Delta$ in the spin paramagnetic energy

$$-\frac{1}{2}\chi_{ij}H_iH_j. \tag{6.49}$$

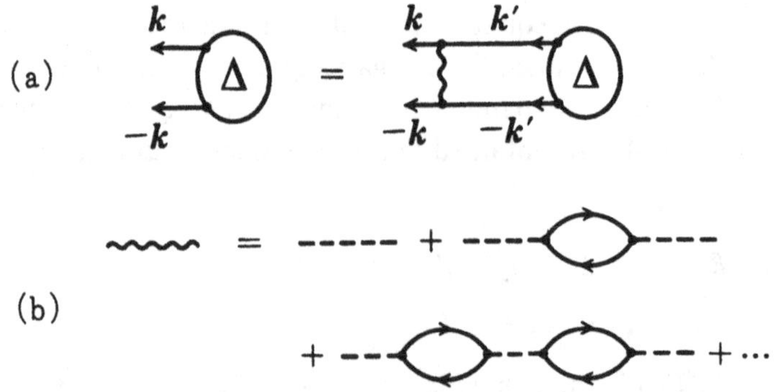

Fig. 6.11. (a) The gap equation. (b) Approximation to the effective interaction (the wavy line).

The magnetic susceptibility in the superfluid state is in general a tensor. One compares Eq. (6.48) with Eq. (6.49) in the BW state to find the constant α_B, noting that if χ_s^0 in Eq. (6.32) is expanded in $\Delta^\dagger\Delta$ at $T \lesssim T_c$, the ratio of the coefficients is given by $\chi_s'/\chi_s^{0'} = (1 + F_0^{(s)})^{-2}$. Then one obtains an additional contribution to Eq. (6.44) of the form

$$f_H = \frac{1}{3}N(0)\Delta_e^2(T)\frac{1}{6}\left(\frac{\gamma}{1+F_0^{(a)}}\right)^2 H_\mu A^*_{\mu i} H_\nu A_{\nu i}. \tag{6.50}$$

Note that, for the magnetic dipole interaction, Eq. (6.35) can be used.

6.5.3 Strong coupling effects and spin fluctuations

It is assumed in the weak coupling theory that the pairing interaction itself is independent of the order parameter. The interaction has been represented by a constant g_1 in the analyses so far. In reality, however, the interaction should change in the presence of pairing since it is an effective interaction in a many-body system including polarisation effects. The wavy line, which represents inter-particle interaction in the gap equation (Fig. 6.11(a)) corresponds to an interaction including polarisation effects, namely density and spin density fluctuations as shown in Fig. 6.11(b). Static polarisation due to an external field has been considered in Section 6.1. Now, one of the interacting quasi-particles plays the rôle of an external field, and hence fluctuations are dependent on space and time. These fluctuations are clearly modified under the superfluid transition. Therefore the gap equation has to be analysed taking the Δ-dependence of the wavy line into account.

Let us consider the gap equation up to the third order in Δ, which

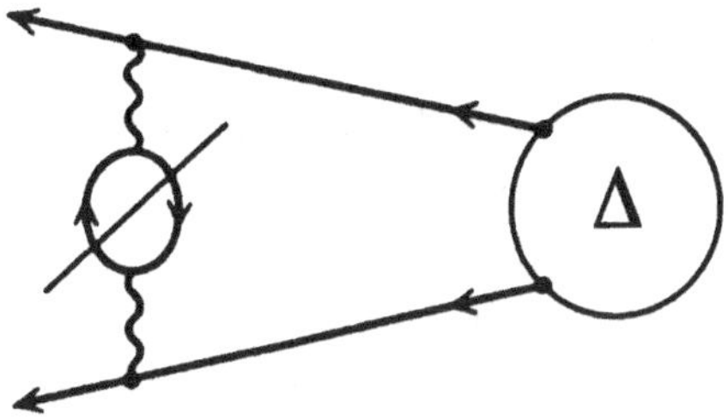

Fig. 6.12. Correction to the coefficients β under the strong coupling theory.

amounts to keeping the fourth order terms in the free energy, since we are interested in the corrections to the coefficients β of the fourth order terms. Terms obtained by expanding the factor $\varepsilon_{k'}^{-1}\tanh(\beta\varepsilon_{k'}/2)$, from the Green's function of the pair $(\boldsymbol{k}', -\boldsymbol{k}')$ in Fig. 6.11(a), in $\Delta^\dagger\Delta$ correspond to the terms with coefficients β already found above. The correction is a term obtained by drawing out $\Delta^\dagger\Delta$, denoted by a slash, from one of the particle–hole pairs in Fig. 6.11(a) and is depicted in Fig. 6.12. The wavy line in Fig. 6.12 is the effective interaction in the normal state. In this discussion only the spin fluctuation, also known as the paramagnon, is considered and the density fluctuation will be neglected since the former is large in liquid ^{3}He as already noted in Section 6.1. Then the wavy line represents $F_0^{(a)}\chi_n$ including the coupling constant with quasi-particles. Here one has to use $F_0^{(a)}\chi_n(\boldsymbol{q}, \omega)$ since fluctuation depends on both the momentum $\boldsymbol{q} = \boldsymbol{k} - \boldsymbol{k}'$ and the frequency. Provided that the ω-dependence is negligible and assuming an appropriate q-dependence, one may estimate the correction $\delta\beta_i$ to the coefficient β_i as

$$\begin{gathered} \delta\beta_1 = \delta\beta_3 = 0, \qquad \delta\beta_2 = -\delta\beta_4 = -\delta\beta_5 = \delta \\ \delta = c(T_c/T_F)[N(0)g_1]^{-2}. \end{gathered} \tag{6.51}$$

Here the coefficient c, which is of the order of unity, increases with pressure, and is a correction stabilising the ABM state at relatively high pressures. Although this model has a resemblance to the attractive force due to phonon exchange studied in Chapter 4, Migdal's theorem is not applicable to spin fluctuations and vertex corrections must be considered to be more precise.

The interaction from spin fluctuations is proportional to the factor $[F_0^{(a)}\chi(q)]^2$, which becomes large as $q \to 0$, as mentioned above. In other words, one may consider physically that ferromagnetic fluctuations are dominant since the Fermi liquid effects enhance the susceptibility. This type of q-dependence favours p-wave pairing.

6.6 Textures and superfluidity

'Symmetry breaking' in superfluid ^4He or an ordinary superconductor has to do with gauge transformations, and the phase χ of the order parameter Ψ was the variable associated with the broken symmetry. It has been shown, moreover, that if the phase χ changes in space a flow of the superfluid with the velocity $\boldsymbol{u}_s(\boldsymbol{x}) = \nabla\chi(\boldsymbol{x})/2m$ appears. On the other hand, symmetries associated with spin and orbital rotations are broken in the superfluid state of ^3He, in addition to that broken under gauge transformation. Because of these facts, there appears in ^3He a superfluid with *texture*, just like a liquid crystal, and its superfluid properties are far more complicated than a conventional superfluid (or a superconductor). These aspects are considered here with emphasis on the relatively simple ABM state.

6.6.1 The London limit

Those states belonging to a particular class, such as the BW state or the ABM state, are mutually connected by 'residual symmetry' transformations. Therefore all the states in a class are energetically degenerate, just as ^1S superconductors with different phase of the order parameter Ψ share common free energy. These considerations show that the most easily realisable among spatial variations should be those involving symmetry transformations within the given class. Therefore one can suppose that the order parameter at an arbitrary point in the system in the B (A) phase belongs to the class of the BW (ABM) state. In the following we will be mainly concerned with spatial variations within this framework. This framework corresponds to the London limit in an ordinary superfluid, which is applicable when the scale of the spatial variation is larger than the coherence length ξ. One may also say that only those variations connected to the Nambu–Goldstone mode are considered in this limit.

6.6.2 Superfluidity in the ABM state

The order parameter in the A phase takes the form of Eq. (6.17), namely $A_{\mu j} = A\hat{d}_\mu(\hat{m}_j + i\hat{n}_j)$, anywhere in the London limit. In other words, one considers a vector $\hat{\boldsymbol{d}}$ and a triad $\hat{\boldsymbol{l}}(= \hat{\boldsymbol{m}} \times \hat{\boldsymbol{n}}), \hat{\boldsymbol{m}}, \hat{\boldsymbol{n}}$. Since they are fields of directed quantities, the superfluid may be considered as if it shows a 'texture'. Of course there appear new characteristics such as anisotropy of physical quantities associated with texture.

The current density (6.47) of a superfluid in the ABM state is written as

$$j_{si} = n_s \left\{ \left(\delta_{ij} - \frac{1}{2}\hat{l}_i\hat{l}_j \right) \hat{m}_k \nabla_j \hat{n}_k + \left(\frac{1}{4}\delta_{ij} - \frac{1}{2}\hat{l}_i\hat{l}_j \right) (\nabla \times \hat{l})_j \right\}, \tag{6.52}$$

where $n_{s\parallel}$ in Eq. (6.29) has been simply written as n_s. When $\hat{\boldsymbol{l}} = \hat{\boldsymbol{m}} \times \hat{\boldsymbol{n}}$ has no spatial dependence, the current j_s depends only on $\hat{m}_k \nabla \hat{n}_k$. If this is the case, $\hat{m}_k \nabla \hat{n}_k$ is just the space-derivative $\nabla \chi$ of the phase χ of the order parameter, as was noted in Section 6.2, and the current has an identical form to that of an ordinary superfluid, except for the anisotropy. Note, however, that the second term does not vanish when $\hat{\boldsymbol{l}}$ varies in space. The $\hat{\boldsymbol{l}}$-vector indicates the direction of the angular momentum of the p-wave pair. Therefore $\nabla \times \hat{\boldsymbol{l}}$ resembles the effective current in a magnet $\nabla \times \boldsymbol{M}$ induced by spatial variation of the magnetisation $\boldsymbol{M}$. The term containing $\nabla \times \hat{\boldsymbol{l}}$, however, contributes to the actual mass current in the present case. This implies that variation in texture leads to superflow since the vector $\hat{\boldsymbol{l}}$ denotes, so to speak, the direction of the texture.

Let us consider, as an actual example, a linear vortex structure which is uniform along the z-axis. Suppose the axis of the vortex is taken to be the z-axis. The uniqueness of the order parameter implies that, although it may change as one moves around the z-axis, it must revert to the same configuration upon return to the starting point. Then the problem is to classify the types of configurations of the field $\hat{\boldsymbol{d}}(\hat{\boldsymbol{m}} + i\hat{\boldsymbol{n}})$ which satisfy the condition mentioned above. Let us suppose, first of all, that the $\hat{\boldsymbol{d}}$-vector is uniform, which may be taken to be $\hat{\boldsymbol{y}}$ for example. One has an ordinary superfluid if $\hat{\boldsymbol{l}} = \hat{\boldsymbol{m}} \times \hat{\boldsymbol{n}}$ is uniform taken to be $\hat{\boldsymbol{z}}$ for example; and the possible structure is a vortex line with a circulation quantum number N(integer). This structure produces the velocity field

$$v_{s\varphi} = N \frac{1}{2m} \frac{1}{r} \tag{6.53}$$

around the z-axis, see Fig. 6.13. The triad rotates around the $\hat{\boldsymbol{l}}$-vector N times as one goes around the z-axis once. This structure has a core of radius $\sim \xi$, within which the superfluid density is suppressed, since v_s diverges as one approaches the z-axis, see Section 5.3. The vorticity $\nabla \times \boldsymbol{v}_s$ is finite only in the vicinity of the core. This type of configuration is called a *vortex structure with a core.*

Let us next consider the case with varying $\hat{\boldsymbol{l}}$. Suppose $\hat{\boldsymbol{l}} \parallel -\hat{\boldsymbol{z}}$ on the z-axis and the vector $\hat{\boldsymbol{l}}$ gradually rises as the distance r from the z-axis increases, eventually satisfying $\hat{\boldsymbol{l}} \parallel \hat{\boldsymbol{z}}$ for sufficiently large r. If $\hat{\boldsymbol{m}}$ and $\hat{\boldsymbol{n}}$ are considered along with $\hat{\boldsymbol{l}}$, one finds that the triad far from the z-axis rotates around $\hat{\boldsymbol{l}}(= \hat{\boldsymbol{z}})$

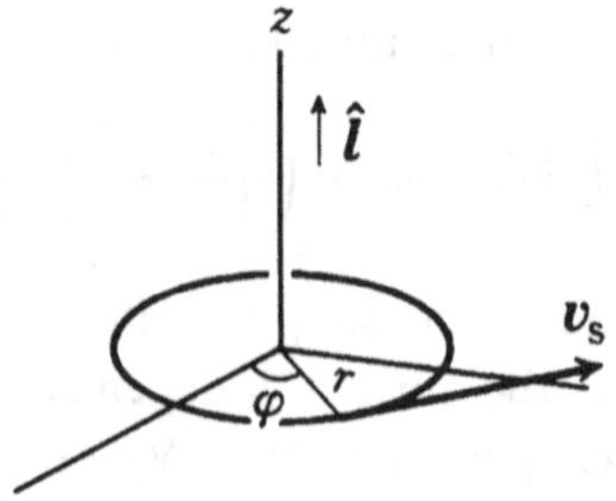

Fig. 6.13. The flow around a vortex line.

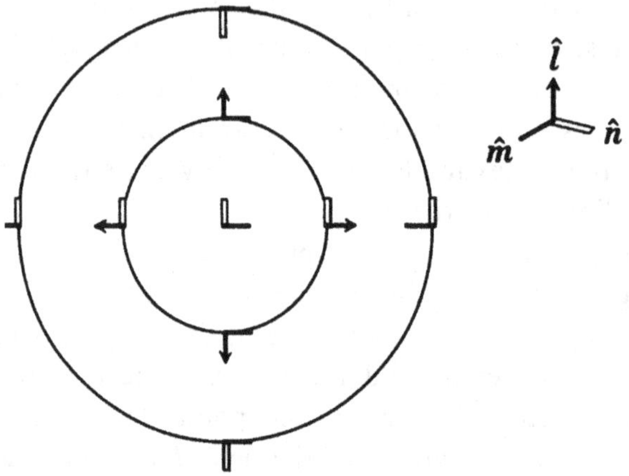

Fig. 6.14. A vortex structure without a core.

exactly twice as one goes around the z-axis once as is shown in Fig. 6.14. Accordingly the structure looks like a vortex with $N = 2$ if seen from far outside, while there is no core associated with the structure since the vortex current vanishes as $r \to 0$. Of course, the circulation is not quantised and the vorticity is finite in a region where $\hat{\boldsymbol{l}}$ varies in space. Provided N is even, a vortex with a core may in general be deformed into such a structure, called the *continuous vortex structure.*

The vector $\hat{\boldsymbol{d}}$ has been kept fixed so far. Now let us consider cases where $\hat{\boldsymbol{d}}$ also varies in space. The order parameter $A\hat{\boldsymbol{d}}(\hat{\boldsymbol{m}} + i\hat{\boldsymbol{n}})$ is left invariant under the simultaneous transformations $\hat{\boldsymbol{m}}, \hat{\boldsymbol{n}} \to -\hat{\boldsymbol{m}}, -\hat{\boldsymbol{n}}$ and $\hat{\boldsymbol{d}} \to -\hat{\boldsymbol{d}}$. Accordingly a vortex structure with a core, as considered previously, with $N = \pm\frac{1}{2}$ is possible, where $\hat{\boldsymbol{m}}$ and $\hat{\boldsymbol{n}}$ rotate around $\hat{\boldsymbol{l}}(=\hat{z})$ by an angle π while $\hat{\boldsymbol{d}}$ reverses in direction as one goes round the z-axis. In other words, the spin $\uparrow\uparrow$ component of the superfluid is in a vortex state while the spin $\downarrow\downarrow$ component is in the $N = 0$ state, or *vice versa*. It should be added that spatial variation of the $\hat{\boldsymbol{d}}$-vector leads in general to a spin current.

6.6.3 Textures

The texture realised under a given condition is determined by minimising the free energy, in particular the gradient energy in the London limit. One must also specify an appropriate boundary condition. Let us first consider the boundary condition imposed on the order parameter at a wall of the vessel. The wall is idealised and considered to be specular and the z-axis is taken to be normal to the wall. Then it is expected that, among three orbital states of the ^{3}P pair, the component $\hat{k}_z$ is destroyed while $\hat{k}_x$ and $\hat{k}_y$ are unaffected, since the z-component of the particle momentum k_z becomes $-k_z$ upon reflection off the wall. Therefore there are no effects of the wall in the ABM state provided that the $\hat{\boldsymbol{l}}$-vector is perpendicular to the wall. If the $\hat{\boldsymbol{l}}$-vector is not perpendicular to the wall, there is a loss of condensation energy of the order of $(F_s - F_n) \cdot \xi(T)$ per unit area. Accordingly the boundary condition at a wall within the London limit is $\hat{\boldsymbol{l}} \parallel \hat{\boldsymbol{n}}$, $\hat{\boldsymbol{n}}$ being the unit normal to the wall. It follows from the same consideration that the London limit is not applicable to the BW state and the planar state is considered to appear in the vicinity of a wall.

The boundary condition $\hat{\boldsymbol{l}} \parallel \hat{\boldsymbol{n}}$ 'pins' the vector $\hat{\boldsymbol{l}}$ perpendicular to the wall in the A phase between parallel plates. In a cylinder, the texture inside the inner circle of Fig. 6.14 is realised in the A phase. The A phase superfluid in a cylinder then automatically supports vortex flow, namely an angular momentum. The magnitude of the vorticity is of the same order as that of an $|N| = 1$ vortex line at the centre.

To determine the texture one must often include the magnetic dipole energy E_D given by Eq. (6.35) in the free energy to be minimised. Similarly one can consider the free energy (6.50) in the presence of an external magnetic field. Where the spatial variation is mild enough, a texture is realised, which minimises these energies everywhere in the sample. In other words, one may apply the London limit not only to the condensation energy but also to other smaller energies as well. The energy scale beyond which the London limit is not applicable is determined from the condition that the gradient energy becomes of the order of the energy under question. If the constant K in Eq. (6.45) is employed in the GL regime, one finds

$$(\xi^2/\xi_D{}^2) \sim \Omega_A^2/\Delta_A^2(T)$$

for the dipole energy and hence the relevant length scale is estimated as

$$\xi_D \sim (\Delta_A(T)/\Omega_A)\xi_0, \tag{6.54}$$

where ξ_0 is the coherence length at $T = 0$. The length scale ξ_D is of the order of 10^{-3} cm according to the above estimate. Similarly the length scale ξ_H defined by the magnetic energy f_H is given by

$$\xi_H \sim (H_A/H)\xi_0, \tag{6.55}$$

where $H_A \sim 2 \times 10^4$ G (= 2 T). The two scales ξ_D and ξ_H are of the same order when $H \sim 28$ G. It follows from Eq. (6.36) that E_D tends to align $\hat{\boldsymbol{d}}$ parallel to $\hat{\boldsymbol{l}}$. It is also found from Eq. (6.50) that an external magnetic field $\boldsymbol{H}$ introduces the condition $\hat{\boldsymbol{d}} \perp \boldsymbol{H}$. The vector $\hat{\boldsymbol{l}}$ satisfies $\hat{\boldsymbol{l}} \perp \boldsymbol{H}$ if these are the only two relevant energies.

6.6.4 Rotating system

One of the most interesting experiments on superfluid ^{3}He is observation of the state realised in liquid ^{3}He rotated together with the vessel. So far system angular velocities of $\Omega \leq 3$ rad/s have been achieved. One expects that a vortex lattice parallel to the rotation axis will appear in this case since coordinate transformation to a rotating system with a constant angular velocity is equivalent to applying a uniform static magnetic field in a charged system. If $N = 1$ vortices appear as in superfluid ^{4}He, the number of vortices per unit area is $N_V = (2m_3/\pi\hbar)\Omega$. Note that the inter-vortex distance r_V is 0.02 cm and $r_V > \xi_D$ for $\Omega = 1$ rad/s. For the above reason, if a magnetic field (> 28 G) is applied parallel to the rotation axis, the vector $\hat{\boldsymbol{l}}$ satisfies $\hat{\boldsymbol{l}} \perp \boldsymbol{H} \parallel \Omega$ outside a region of radius ξ_D from the centre of each vortex, as shown in Fig. 6.15. The vector $\hat{\boldsymbol{l}}$ varies in the region $r < \xi_D$ and the vorticity is expected to be finite there. Figure 6.16 shows one candidate vortex structure that minimises the free energy. One technique to study the texture experimentally is nuclear magnetic resonance. A satellite appears in the resonance absorption spectrum with area proportional to the angular velocity, that is, the number of vortices in the structure when the system is under rotation, see Fig. 6.17. The vector $\hat{\boldsymbol{l}}$ is unlocked from $\hat{\boldsymbol{d}}$ for $r < \xi_D$, which leads to a potential energy due to E_D. Then a localised spin wave mode appears which is considered to be responsible for the satellite. These facts support the proposed continuous vortex structure.

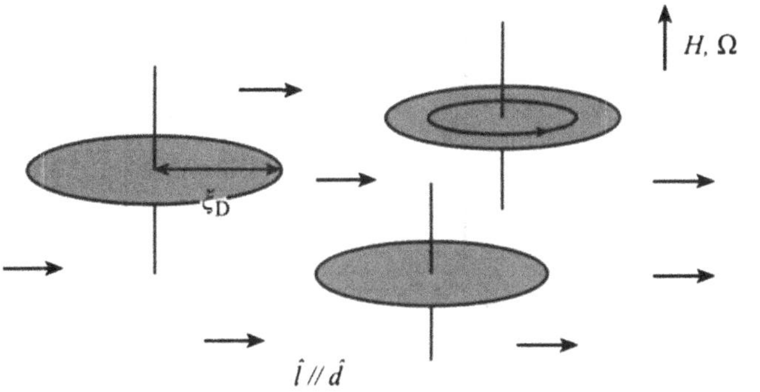

Fig. 6.15. The $\hat{l}$-vector satisfies $\hat{\boldsymbol{l}} \parallel \hat{\boldsymbol{d}} \perp \boldsymbol{H}, \boldsymbol{\Omega}$ outside the shaded regions.

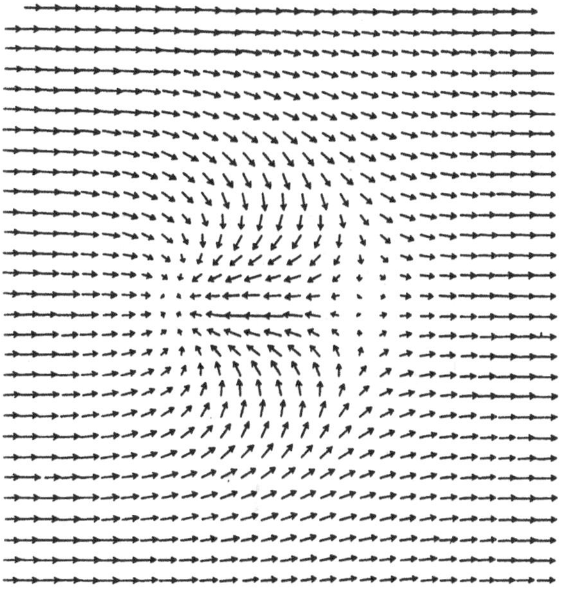

Fig. 6.16. The $\hat{l}$-texture associated with vortices.

6.6.5 Vortex in B phase

So far vortices in the ABM state have been considered. In fact vortex structures in the B phase are also quite interesting. Figure 6.18 shows the NMR data obtained from rotating superfluid in the B phase. There are many peaks due to the spin waves produced by the $\hat{\boldsymbol{n}}$-texture when the B phase fluid is under rotation. The vortices produced by rotation lead to an increase in the interval between two neighbouring peaks as shown in Fig. 6.18. These resonance frequencies depend on the temperature and the pressure, and jump

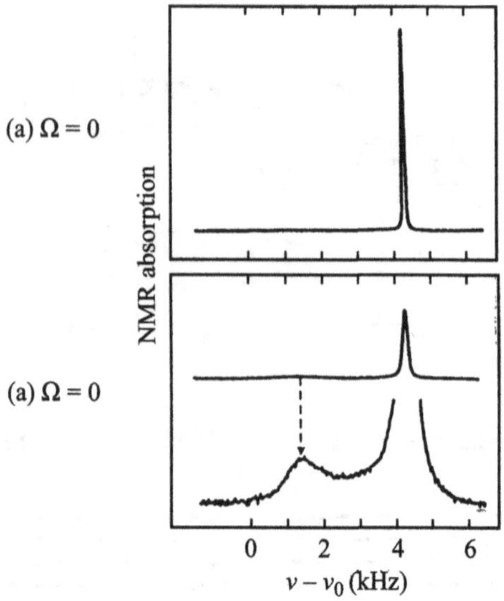

Fig. 6.17. Appearance of a satellite in the NMR absorption curve when liquid ^{3}He in the A phase is rotated. (From [G-20].)

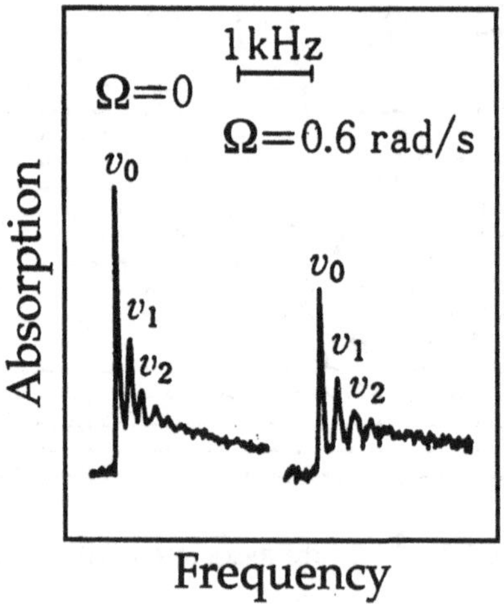

Fig. 6.18. The change in the NMR absorption curve when liquid ^{3}He in the B phase is under rotation. (From [G-20].)

suddenly at some temperature as shown in Fig. 6.19. This demonstrates that in the B phase there are two phases of a vortex structure, V_A and V_{NA} as shown in Fig. 6.20. Moreover it is found that the resonance frequency of the low temperature phase depends on whether $\boldsymbol{\Omega} \parallel \boldsymbol{H}$ or $\boldsymbol{\Omega} \parallel -\boldsymbol{H}$, hence spontaneous magnetisation is associated with the vortex structure.

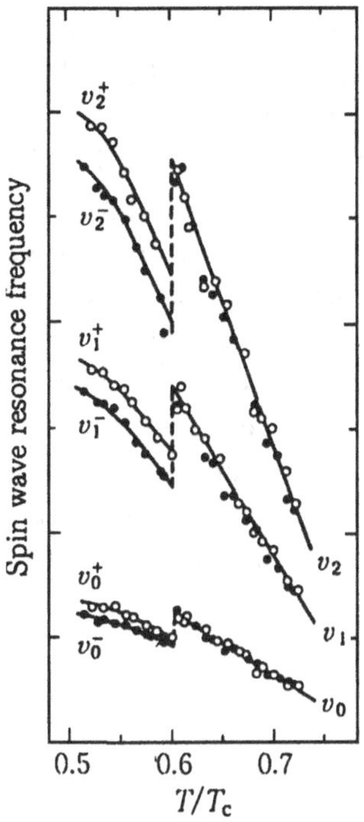

Fig. 6.19. The temperature dependence of the NMR frequency due to spin waves. (From [G-20].)

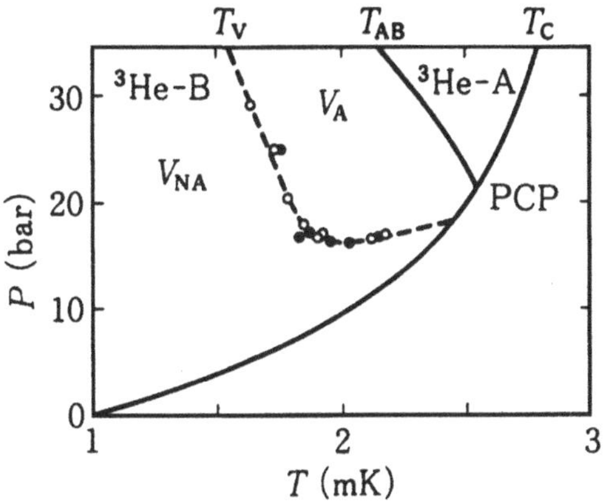

Fig. 6.20. The phase diagram of a vortex structure in the B phase. The region V_A denotes the phase with an axially symmetric vortex while V_{NA} denotes the phase with a vortex without axial symmetry. (From [G-21].)

The BW state has an isotropic gap and behaves in a similar way to an ordinary superfluid in the London limit. Accordingly the vortex structure is no different from that of a quantised vortex line with a core if seen from a distance $r \gg \xi$ from the centre. The London limit is not applicable, however, in the vicinity of the core $r \lesssim \xi$ and there appears a structure which deviates from the BW state. To analyse this structure theoretically, one minimises the GL free energy up to the fourth order terms with such boundary conditions that the structure approaches the vortex in the BW state at $r \gg \xi$. According to theoretical calculations, the core part is filled with a superfluid similar to the ABM state. There is an axially symmetric vortex V_A and asymmetric vortex V_{NA} and the latter is the candidate for the structure that appears in the low temperature (or low pressure) phase. The core of an asymmetric vortex can be regarded as made up of half-quantised vortex pairs in the A phase.

7
New superconducting materials

Ever since the discovery of the phenomenon by Onnes, the quest for new superconducting materials has enlivened the history of research in superconductivity. The strongest motivation for this research has been the search for a superconductor with higher T_c. Since the discovery of superfluid ^{3}He, non-s-wave superconductors have also been sought. The fruit of the former is represented by discovery of the high temperature superconductors (abbreviated hereafter as HTSC) in 1986 while the heavy electron system (also called the heavy fermion system) is highlighted as an example of the latter. The characteristics of these two systems are discussed in the present chapter.

7.1 Superconducting materials

Let us first outline typical materials undergoing superconducting transition. Table 7.1 shows characteristic materials from several categories. It was shown in Chapter 4 that, where the electron–phonon interaction is responsible for the superconductivity, $T_c \sim \omega_D \exp\left(-\frac{1+\lambda}{\lambda-\mu^*}\right)$ (or Eq. (4.48) to be more precise). Accordingly the system should become a superconductor if the parameter λ for the attractive interaction exceeds the effective Coulomb interaction parameter μ^*, which is about 0.1 in an ordinary metal. Since large λ implies a large electrical resistivity, a bad conductor as a normal metal tends to be a superconductor. In fact, all the metallic elements, except magnetic metals such as Co, Ni, Fe, alkaline metals and noble metals, become superconducting at low enough temperatures including such elements as Si, S and I that undergo metallic transition under high pressure.

The parameter λ is large if the density of states $N(0)$ on the Fermi surface and the electron–phonon coupling constant are large. Metals with a rather small Debye frequency are strongly coupled. Figure 7.1 shows that there is a correlation between $N(0)$ [$= \frac{3}{2}\gamma/(\pi k_B^2)$] and T_c. There have been

Table 7.1. *Various superconductors*

	Material	T_c (K)	
Elements	Au	< 0.001	$\lambda \sim 0.1$. Weak electron–phonon interaction.
	Al	1.2	$\lambda \sim 0.4$. Typical BCS type.
	Pb	7.2	$\lambda \sim 1.55$. Strong coupling system.
	Nb	9.2	$\lambda \sim 1$. A d-band metal. The element with the highest T_c.
	Ga	8.6	$T_c = 1.1$ K for an amorphous crystal.
Under pressure	Si	7	A metal at $P > 1.2 \times 10^{10}$ Pa.
	I	1.2	A metal at $P > 2.1 \times 10^{10}$ Pa.
	(Metallic H	$135 \sim 200$?	Large ω_D. A metal at $P > 1.5 \times 10^{11}$ Pa.)
Metallic compounds and alloys	Nb_3Sn	18	A15 type, $\lambda \sim 1.65-1.95$.
	Nb_3Ge	23	Thin film, the largest T_c among A15 type.
	$Gd_{0.2}PbMo_6S_8$	14.3	Chevrel type, the highest H_{c2} except for HTSC.
	$ErRh_4B_4$	8.5	Becomes ferromagnetic at 0.9 K, superconductivity disappears.
	$ErMo_6S_8$	2.2	Superconductivity coexists with antiferromagnetism at $T < 0.2$ K.
Heavy electron systems	$CeCu_2Si_2$	$0.5 \sim 0.6$	
	UPt_3	0.5	Three distinct superconducting phases, ^{3}P-pairing(?).
	UBe_{13}		
Oxides	$SrTiO_{3-x}$	$0.05 \sim 0.5$	A superconductor with a small carrier density $n < 10^{20}$ cm^{-3}.
	$LiTi_2O_4$	13	Perovskite structure.
	$Ba(Pb_{1-x}Bi_x)O_3$	13	$x \sim 0.25$. An insulator at $x > 0.35$.
	$Ba_{1-x}K_xBiO_3$	> 30	$x \sim 0.4$, an electron–phonon type (?), three-dimensional.
Copper oxides	$La_{2-x}Ba_xCuO_4$	> 30	The first copper oxide, two-dimensional.
	$YBa_2Cu_3O_{7-\delta}$	93	
	$Bi_2Sr_2CaCu_2O_8$	125	
Organic compounds	$(TMTSF)_2ClO_4$	$1.2 \sim 1.4$	One-dimensional.
	$[BEDT{\cdot}TTF]_2{\cdot}Cu(NCS)_2$	11.4	The largest T_c among organic materials, two-dimensional.
	K_3C_{60}	~ 18	Doped fullerene, $\lambda \sim 0.5$ (?).
	Rb_3C_{60}	~ 28	

Remarks: A table based on data around 1980 (P.B. Allen and B. Mitrovič [E-3]) was used in compiling the above table. The data for I are taken from recent research. (From [G-22].)

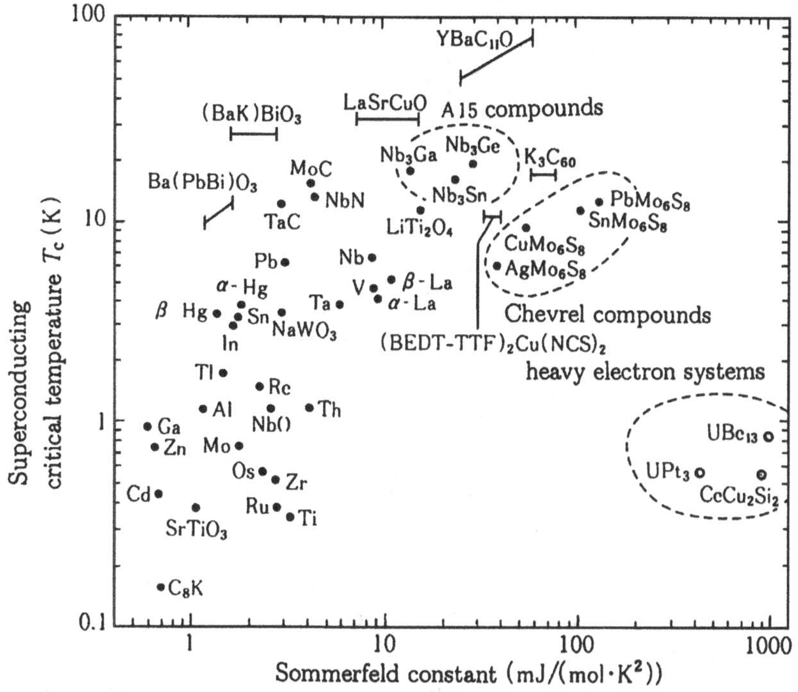

Fig. 7.1. The relation between the Sommerfeld constant γ and T_c.

numerous theoretical works, based on the strong-coupling theory, on how far T_c may be raised when ordinary electron–phonon interaction is responsible for superconductivity. Although no definite conclusion has been reached, it has been considered that the upper bound of T_c is about 30 $\sim$ 40 K since very large λ may lead to lattice deformation.

Another approach towards high T_c is to look for an inter-electron force in mechanisms other than phonon exchange. The exchange of bosons, not necessarily phonons, with an appropriate spectrum leads to an attractive force between electrons. One example of this mechanism is the case of liquid ^{3}He mentioned in Chapter 6, where a part of the attractive force is due to spin density fluctuation called the paramagnon. The first proposal along this line was the excitonic mechanism. An example of this mechanism is depicted in Fig. 7.2, in which an electron moving along the primary chain polarises the side branches which in turn attract another electron to result in an attractive force between these electrons, see [G-23]. It is expected that this system might have large T_c since, instead of the Debye frequency ω_D, a far larger exciton energy appears as a prefactor of the exponential function in the equation for T_c. Although such a proposal suffers from theoretical

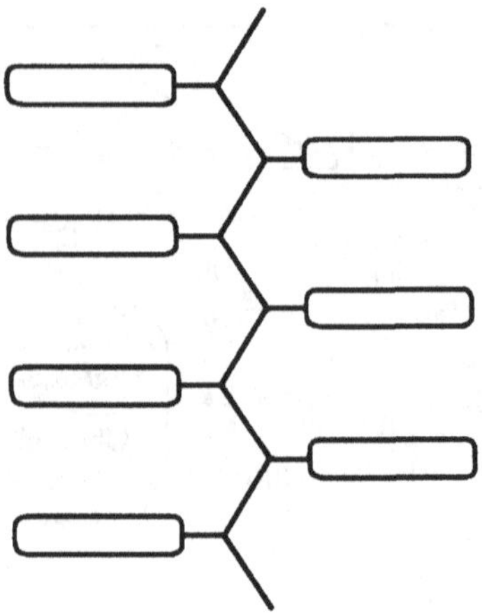

Fig. 7.2. A model of an exciton mechanism.

difficulties and has had no specific successes, it has stimulated research in organic conductors and many organic conductors showing superconductivity have now been fabricated.

The quest for new superconducting materials has in reality proceeded empirically. The first big success was the discovery of the A15 superconductors which followed the discovery of the Chevrel metallic compounds with large H_{c2}. One of the characteristics of the A15 superconductors is that transition metals with d-electrons line up along x-, y- and z-directions like chains as shown in Fig. 7.3. Present understanding from the tunnelling spectrum and other means is that superconductivity in the A15 system is due to electron–phonon interaction. It should be remembered that there is a structural phase transition competing with superconductivity in the A15 system.

The next remarkable progress is the discovery of $BaPb_{1-x}Bi_xO_3$ by A.W. Sleight, J.L. Gillson and P.E. Bierstedt ([G-24]) which is the first system showing superconductivity among oxides basically of the perovskite type, see Table 7.1. $BaPbO_3$ is a semi-metal while $BaBiO_3$ is an insulator of the Peierls type. A metallic phase showing superconductivity appears between these two systems. A member of the same group, $Ba_{1-x}K_xBiO_3$ with T_c as large as $\gtrsim 30$ K was found later. From recent research it appears that their properties might be explained within the framework of electron–phonon interaction.

The research on copper oxide high-T_c superconductors which pushed up T_c from 23 K (Nb_3Ge) to 120 K at a stroke was triggered by the discovery of $La_{2-x}Ba_xCuO_4$ in 1986 ([G-1]). The cuprates have not only large T_c but other unique properties described in the next section.

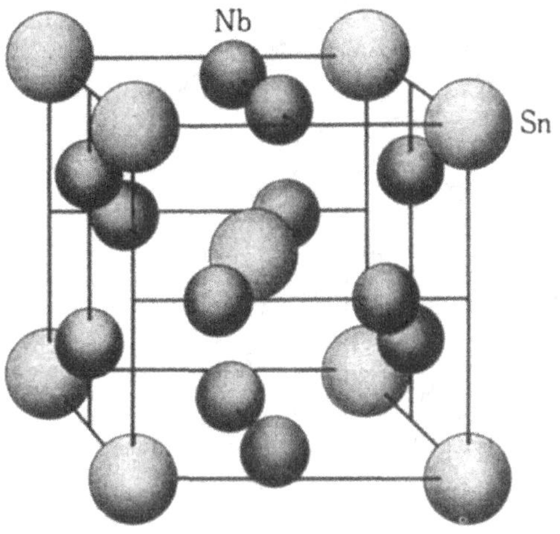

Fig. 7.3. The crystal structure of Nb_3Sn (A15 type).

Let us add a few lines on organic superconductors before we close this section. Typical systems in this group include the TMTSF system ($T_c = 0.4 \sim 1.4$ K) and $(BEDT\text{-}TTF)_2X$ ($T_c = 10 \sim 12.8$ K) in Table 7.1. Figure 7.4 shows the crystal structure of $(BBDT\text{-}TTF)_2[Cu(NCS)_2]$. Their common property is conductivity of the charge-transfer type with small carrier density ($n \sim 10^{21}$ cm^{-3}) and low dimensionality. For the compound shown in Fig. 7.4, the coherence length in the bc-plane is $\xi_{\parallel} \sim 30$ Å while that along the a-axis is $\xi_{\perp} \sim 3$ Å. The nesting of the Fermi surface becomes important in these systems due to low dimensionality so that a state with a charge density wave or a spin density wave can appear under pressure for example. It is not yet established whether the superconductivity is due to s-wave pairs or not. An interesting system related to this is C_{60}, one of the fullerenes, doped with an alkaline metal K or Rb, see Fig. 7.5. According to popular theory the conduction electrons in this system are mainly π-electrons and their coupling with the intra-molecular vibration of C_{60}, whose energy is $1000 \sim 2000$ K, is responsible for rather large T_c.

7.2 Copper oxide superconductors

7.2.1 Structure

There are more than ten copper oxide superconductors (abbreviated as HTSC hereafter) known to date. Structures common to these are the CuO_2 sheets in an alternating layer structure with blocks of a metallic element other than Cu (alkaline earth, rare earth, Bi, Tl and so on) and oxygen.

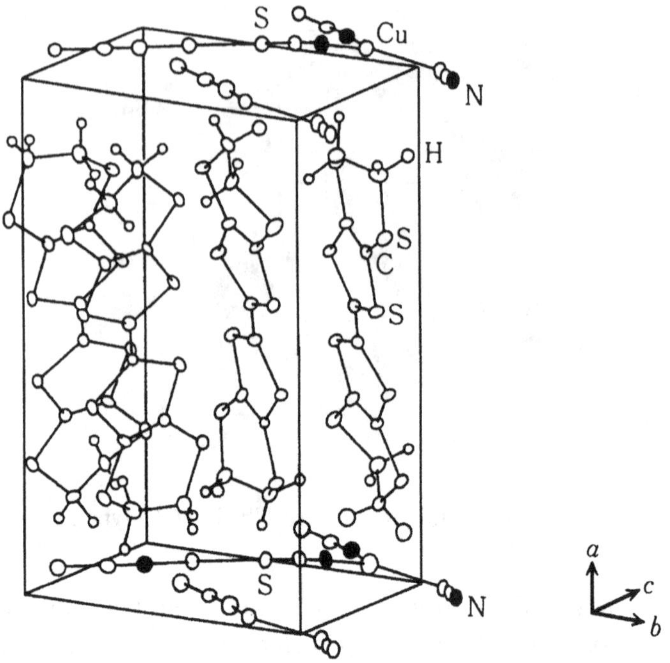

Fig. 7.4. The crystal structure of $(BEDT\text{-}TTF)_2[Cu(NCS)_2]$.

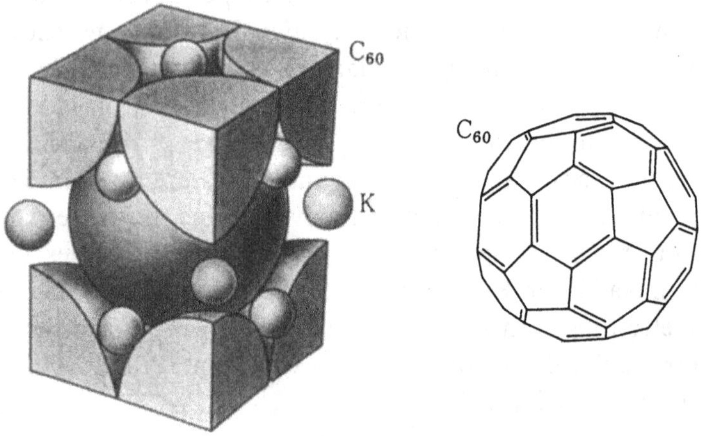

Fig. 7.5. The crystal structure of K_3C_{60}.

Four typical HTSC structures are shown in Fig. 7.6, where the CuO_2 sheet is seen to be sandwiched by the shaded block layers. There are two kinds of CuO_2 sheet, one with a Cu atom surrounded by O atoms at the corners of an octahedron or a pyramid and the other with all the atoms in a sheet as

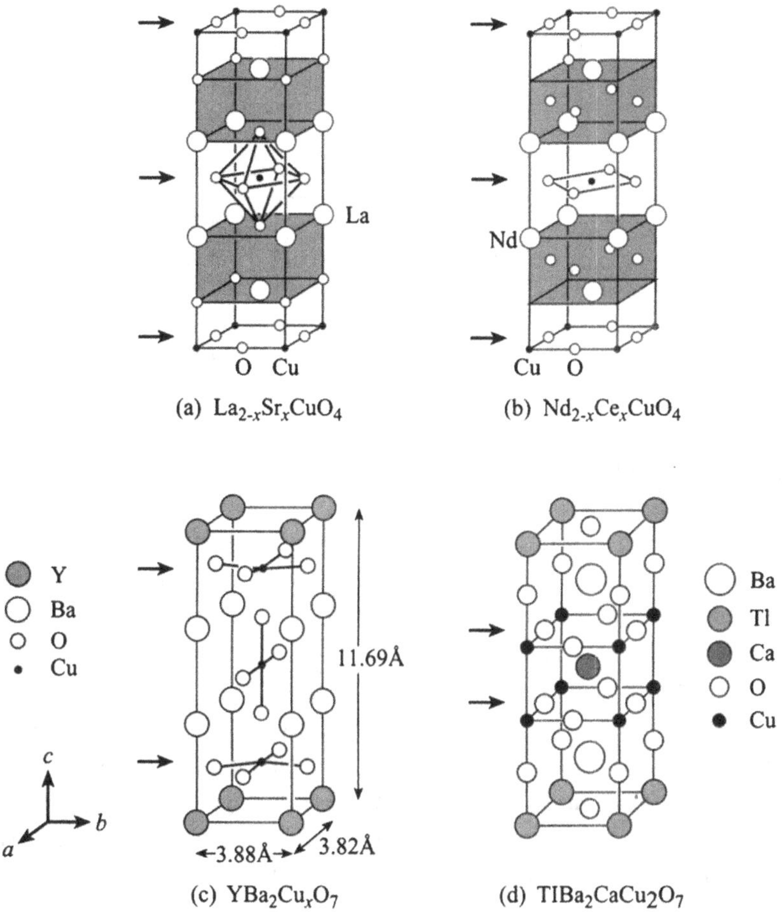

Fig. 7.6. Typical crystal structures of HTSC. The arrows indicate the CuO_2 sheets. (See Yoshinori Tokura in [D-16] for the structure.)

in Fig. 7.6(b). Properties such as doping differ between these sheets but we will not be concerned with that difference here.

7.2.2 Electronic structure

In order to understand the basic electronic states of HTSC such as $La_{2-x}Ba_xCuO_4$ or $La_{2-x}Sr_xCuO_4$ (Fig. 7.6(a)), discovered first, it is preferable to study their parent material La_2CuO_4. Two electrons are transferred from the La_2O_2 layer to the CuO_2 sheet; thus on average the system is charged as $(La_2O_2)^{2+}(C\,O_2)^{2-}$. Accordingly in this crystal the coupling between the two layers is ionic in nature. Then the 3d shell of Cu has a

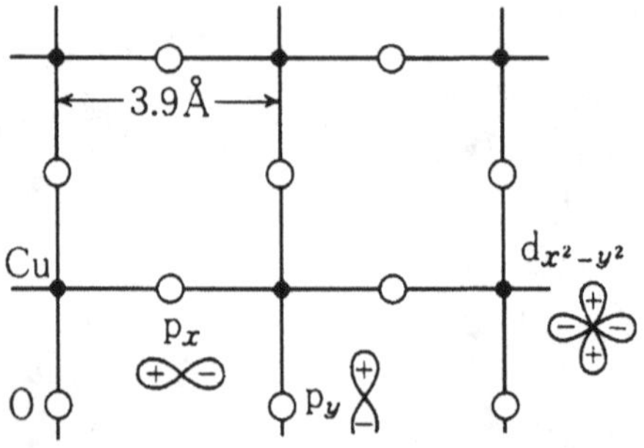

Fig. 7.7. The orbitals used in the three-band model of the CuO_2 sheet.

single hole while the 2p shell of O is closed as far as the formal valency is concerned.

Let us examine in greater detail the electronic structure of the CuO_2 sheet, which plays the central rôle in what follows. The 4s orbitals of Cu may be neglected since they form a broad band. The 3d orbitals are relevant to the present problem. It is important to realise that, compared to other 3d transition metals, the energy of the 3d orbitals of Cu is the closest to that of the 2p orbitals of O. Accordingly these two orbitals tend to hybridise. Suppose the CuO_2 sheet is taken to be the xy-plane. Among the 3d orbitals $d_{x^2-y^2}$ has the largest amplitude in the plane and, together with p_x and p_y of O, makes up three hybridised orbitals (Fig. 7.7) having highest energy and accomodating five electrons. Of the three bands, the one with the largest energy is half-filled from the band theoretical viewpoint. Hence we expect this system to be metallic. Actually La_2CuO_4 is an insulator and, moreover, is antiferromagnetic below $\sim$ 300 K. The reason for this behaviour is found in the strong Coulomb interaction between electrons. In particular, the 3d orbitals of Cu have a small extension in the crystal and the Coulomb energy U, when two electrons (spin $\uparrow$ and $\downarrow$) occupy the $d_{x^2-y^2}$ orbital of the same site, is larger than the band width. Thus the system is a *Mott insulator* when each site is occupied by exactly one electron. It is well known from discussion on the Hubbard model that in such a case the spins align in an antiferromagnetic way at low temperatures due to exchange interaction, see [A-6].

Suppose a part x of La is replaced (i.e., doped) by Ba or Sr. Then the number of electrons transferred to CuO_2 decreases by x, thus there appear holes with concentration x in the Mott insulator. (It is known from the light absorption data that holes mainly exist near the O atoms.) These holes move around the crystal without being blocked by the Coulomb repulsion

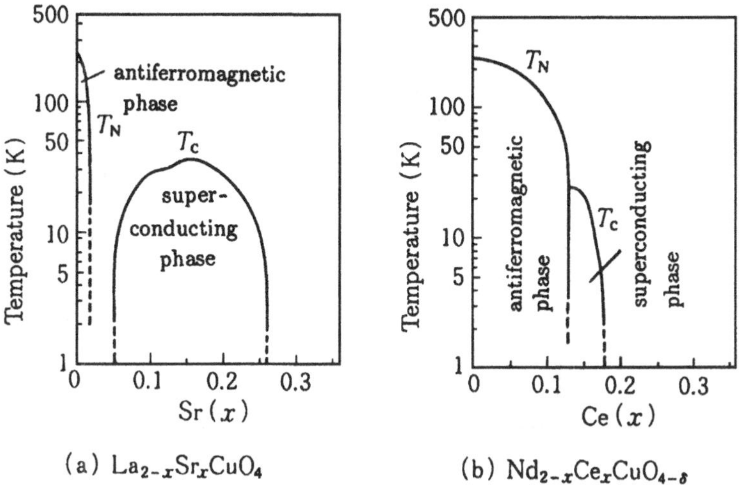

Fig. 7.8. Phase diagrams of (a) the La system and (b) the Nd system. T_N is the Néel temperature at which the system undergoes a phase transition to the antiferromagnetic phase. (From Yasuo Endoh in [D-16].)

U and hence the system becomes metallic as x is increased. A similar change appears in $Nd_{2-x}Ce_xCuO_4$ in Fig. 7.6(b) for which even more electrons are supplied by doping. Note that the superconducting phase is adjacent to the antiferromagnetic phase as found from the low temperature phase diagram (Fig. 7.8).

An extended three-band Hubbard model with the Hamiltonian

$$H = \sum(\varepsilon_d d^\dagger d + \varepsilon_p p^\dagger p) - t_{pd}\sum(d^\dagger p + p^\dagger d) - t_{pp}\sum p^\dagger p$$
$$+U_d\sum n_{d\uparrow}n_{d\downarrow} + U_p\sum n_{p\uparrow}n_{p\downarrow} \tag{7.1}$$

is used in the analysis of the many-body problem in an HTSC system. Here $d^\dagger, d$ and $p^\dagger, p$ are the sets of creation and annihilation operators of electrons in the d orbital of the Cu atom and the p orbital of the O atom respectively, where the site index has been omitted. Rough estimates of the important parameters are $\varepsilon_p - \varepsilon_d \sim 2.5$ eV, $t_{pd} \sim 1$ eV, $t_{pp} \sim 0.5$ eV, $U_d \sim 10$ eV and $U_p \sim 4$ eV.

Suppose the system enters a metallic state leading to superconductivity as carrier density is increased by doping. Then questions to ask are whether the Fermi liquid theory is applicable or if new kinds of elementary excitations have to be introduced due to strong correlations; and what is the mechanism for superconductivity? Since there seem to be no definite answers to the above questions at present, let us next summarise the observed characteristic properties of HTSC in the superconducting state.

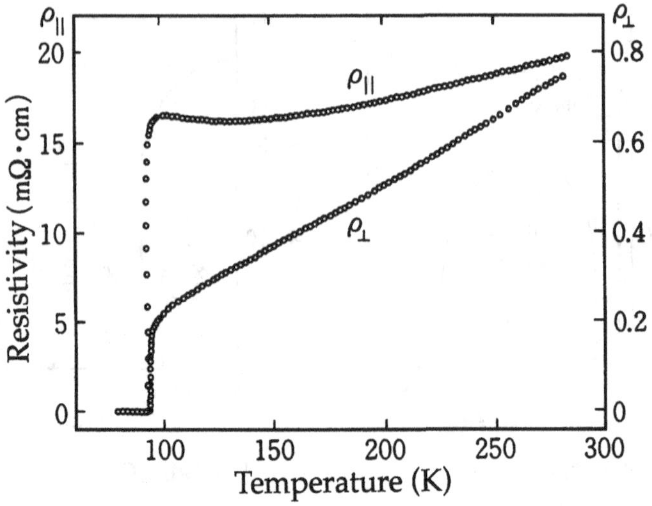

Fig. 7.9. The resistivity of YBCO at zero magnetic field. $\rho_{\parallel}$ denotes the resistivity along the c-axis while $\rho_{\perp}$ denotes that in the ab-plane. (From [G-25].)

7.2.3 Superconducting state

1. **Electrical resistivity** Figure 7.9 shows the temperature dependence of the electrical resistance of $YBa_2Cu_3O_7$ (YBCO) with no magnetic field. The resistivity in the ab-plane (CuO_2-plane) is quite different from that along the c-axis perpendicular to the plane as expected from the layered structure described above. In other words, the system is two-dimensional so far as electrical conduction is concerned. The two-dimensionality is even stronger in Bi systems. It must be noted next that the resistivity ρ in the normal state is proportional to T as $\rho = aT$ over a broad temperature range. Deviation from this behaviour should be found if the resistivity is due to scattering by phonons, though there are no quantitative calculations on this yet. Finally it should be added that the Hall coefficient R_H in the normal state shows a notable temperature dependence, which cannot be explained from the ordinary Fermi liquid result $R_H = 1/ne$ where n is the carrier number density.
2. **Cooper pair** It is certain from measurement of the magnetic flux quantum ($hc/2e$) and the observation of the Shapiro steps (see Section 3.4) that pairing of electrons is responsible for the superconductivity in HTSC. Whether the pair is of s-wave or non-s-wave type has not yet been determined.
3. **Coherence length** The coherence length ξ at $T = 0$ (the subscript

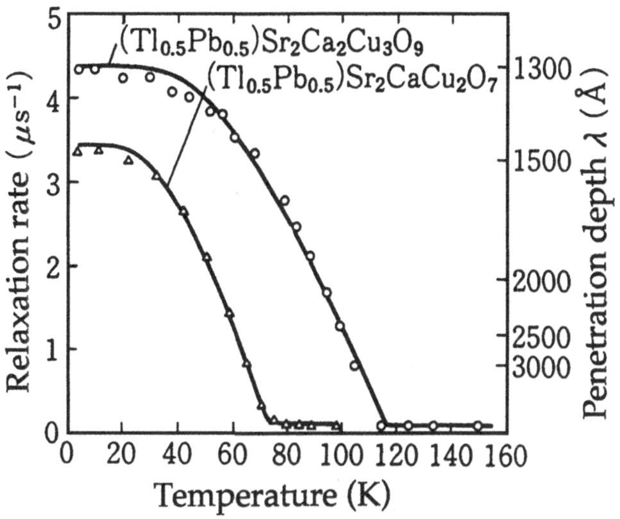

Fig. 7.10. The penetration depth of two HTSCs obtained from muon spin relaxation. The solid lines are the prediction of BCS theory. (From [G-26].)

0 is omitted) is extremely small and has a strong anisotropy. If $\xi_a \sim \xi_b \sim 15$ Å is employed in the CuO_2-plane, there are only 4×4 CuO_2 unit cells in the area ξ_a^2. Put another way, provided that each unit cell delivers a single carrier, the ratio of ξ_a to the average inter-carrier distance is 4, which is quite different from the ratio $10^3 \sim 10^4$ for an ordinary superconductor. The coherence length ξ_c along the c-axis is about 2 Å even for YBCO with relatively small anisotropy. Thus the interlayer coupling is rather close to a Josephson junction and, when discussing the microscopic mechanisms of superconductivity, we can regard the pair wave function to be confined in the CuO_2-plane. Note that the finiteness of the mean free path l may be negligible (the clean limit) since ξ is so small that $l \gg \xi$.

4. **London penetration depth λ_L** The length λ_L is extremely large in contrast with ξ. The most reliable measurement seems to be the muon spin relaxation (μSR) result, which makes use of de-phasing in the muon spin precession induced by an inhomogeneous magnetic field. Figure 7.10 shows the temperature dependence of λ_L measured in the flux lattice state of YBCO with $\boldsymbol{H} \parallel \hat{c}$, which roughly reproduces the temperature dependence of the superconducting component predicted by the BCS theory if $\lambda_L^{-2} \propto n_s(T)$ is assumed. Since the GL parameter $r = \lambda/\xi$ is ~ 100, HTSC is a typical 'type 2 like' superconductor.

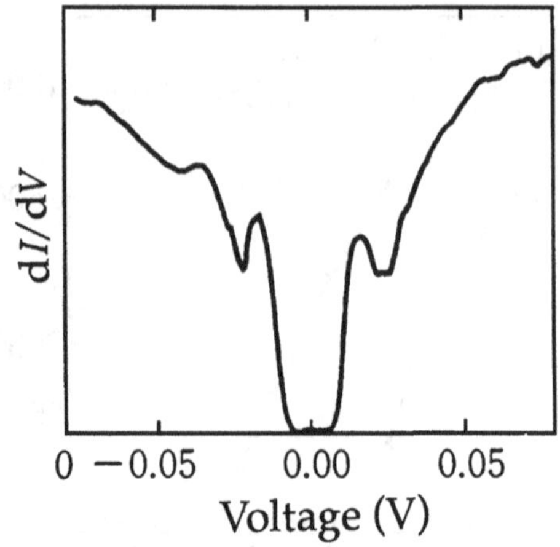

Fig. 7.11. The gap structure of a thin film of YBCO observed by STM at 4.2 K. (From [G-27].)

5. **Energy gap** Δ The energy gap may vanish at points or along lines on the Fermi surface, as found in Chapter 6, if the pairing is of non-s-wave type. Then at low temperatures the thermodynamic quantities and relaxation time approach zero as a power of T, rather than $\exp(-\beta\Delta)$. Even for s-wave pairing one must keep in mind that the temperature dependence of Δ may deviate from the BCS result in the strong coupling regime. Although the energy gap Δ may be directly observed in the electromagnetic absorption spectrum, tunnelling measurements and so on, no definite data are yet available partly due to the smallness of ξ. According to the recent tunnelling conductance measurement using scanning tunnelling microscopy (STM) (Fig. 7.11), there are no excitations within the gap, which favours s-wave pairing. The ratio $\Delta_0/k_B T_c$ for all HTSC materials seems to be much larger than the BCS value given in Eq. (3.14).

6. **Nuclear spin relaxation time** T_1 The inverse of the nuclear spin relaxation time, T_1^{-1}, of an s-wave superconductor of the BCS type is peaked at just below T_c and is exponentially small at low temperatures due to the coherence factor and the energy dependence of the density of states as shown in Chapter 3. Figure 7.12 shows the T_1 of ^{63}Cu in HTSC. Observe that T_1^{-1} has no peak at just below T_c and does not decay exponentially at low temperatures. Although this fact may be explained if d-wave pairing is assumed, it is not clear if the s-wave pairing should definitely be rejected when such effects as depairing

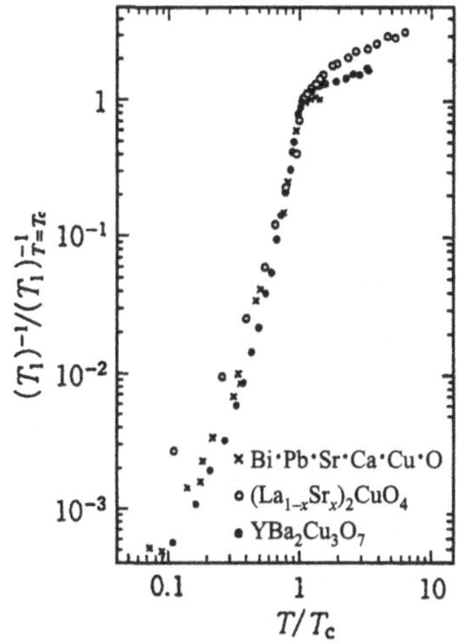

Fig. 7.12. The nuclear spin relaxation rate, normalised by the value at T_c, measured from the nuclear quadrupole resonance of ^{63}Cu. (From [G-28].)

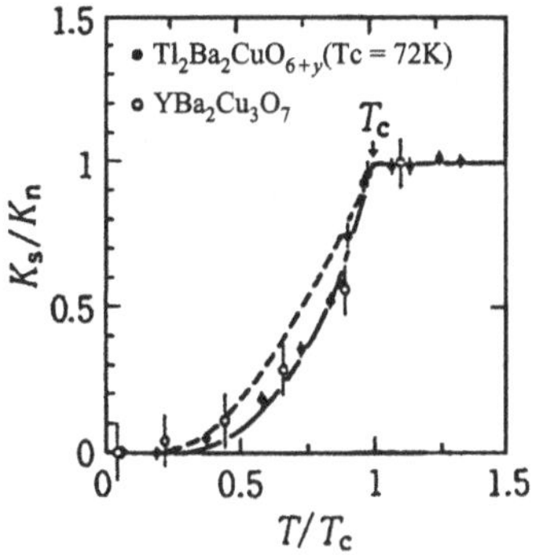

Fig. 7.13. The temperature dependence of the ratio (K_s/K_n) of the Knight shifts. The residual shift at $T = 0$ is subtracted from the data. The dotted (solid) line depicts the result of BCS theory with $2\Delta = 3.5\,(4.5)\,k_B T_c$. (From [G-28].)

are considered. The Knight shift measurement is given *en passant* in Fig. 7.13. The approximate agreement between Fig. 7.13 and the BCS result obtained in Section 3.3 should not be taken seriously since the shift at $T = 0$ is assumed to be totally from the orbital part and has been subtracted from the data in Fig. 7.13.

7. **Neutron diffraction** Since La_2CuO_4 shows antiferromagnetism, it is considered that the antiferromagnetic spin fluctuation remains strong even when the system is doped to be a metal. In fact this fluctuation has been observed in inelastic neutron scattering experiments.

Interesting data such as the temperature dependence of the phonon spectrum density are currently being obtained.

7.2.4 Mechanism of superconductivity

There has been intensive theoretical research on the mechanism of superconductivity in HTSC based on the three-band Hubbard model (7.1) or its more simplified variant called the t–J model. It is interesting to ask what kind of long range order appears in the ground state when the filling factor deviates from 1/2, independently of its relevance to the actual HTSC. Although there are many attractive attempts to determine this, such as the resonating valence band theory and the anyon theory, so far none seems to yield consistent physical results. (See [D-12], [D-13] and [D-14] for reviews of recent research.) An attempt along a conventional line of arguments is a theory which makes use of bosonic antiferromagnetic fluctuations, rather than phonons. Since these fluctuations become attractive at large wave numbers, in contrast with paramagnons, they can lead to d-wave pairing, see Fig. 2.2. It is reported that resistance proportional to T may be obtained in the normal state when electrons are scattered mainly by antiferromagnetic fluctuations. On the other hand, phonons may not be totally negligible since HTSC, as with A15 systems, is always accompanied by lattice instability. The central questions are what plays the main rôle in the pairing and what type of pair appears as a result?

7.2.5 HTSC in a magnetic field

Some of the remarkable features of HTSC are small coherence length and quasi-two-dimensionality as was noted in Section 7.2.3 and they result in new phenomena that have not been observed in other superconductors. They are particularly noticable in HTSC in a strong magnetic field.

1. **Fluctuations** The smallness of the coherence length ξ implies that the effect of fluctuations will be substantial. In fact, Fig. 7.9 shows that the resistivity starts to decrease well above T_c. This is the result of the superconducting fluctuations treated in Section 5.7 (see Eq. (5.58a)) for $T \gtrsim T_c$, which can be observed in bulk HTSC, not just in thin films.

   The most remarkable features of the system are observed in a magnetic field. Figure 7.14 shows the temperature dependence of the electrical resistivity of a YBCO single crystal in a magnetic field. As

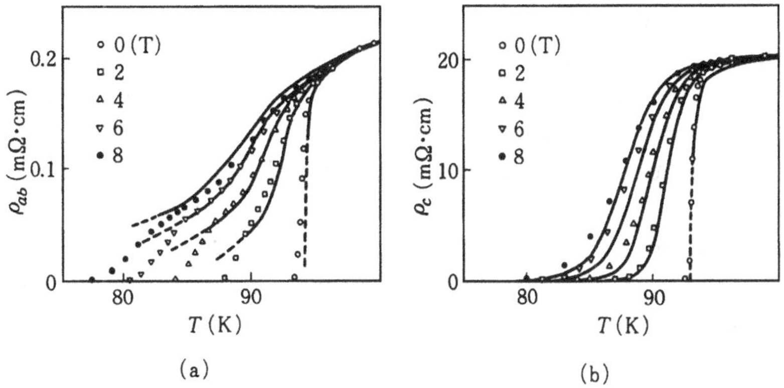

Fig. 7.14. The temperature dependence of the resistivity along the *c*-axis. (a) The current I is in the *ab*-plane. (b) $I \parallel c$. The solid lines are due to a theory with non-Gaussian fluctuations. (Observed values are taken from [G-29]. Theoretical curves are from T. Tsuneto, R. Ikeda and T. Ohmi in [D-16].)

B is increased, the curve $\rho(T)$ of an ordinary type 2 superconductor should shift towards the low temperature side as T_{c0} shifts to $T_{cH} = T_{c0}(1 - \phi_0/B\xi_0^2)$ (a relation obtained by solving Eq. (5.29) for T_c). In contrast with this, Fig. 7.14 clearly shows different behaviour, which makes it impossible to define $H_{c2}(T)$. In other words, the transition from the normal state to the Abrikosov flux lattice state mentioned in Chapter 5 is not clear-cut in the present case.

We can use the Ginzburg criterion introduced at the end of Section 5.7 to see whether or not fluctuations are of essential importance in the present system. The critical region in the absence of a magnetic field is $|\varepsilon| \equiv |1 - T/T_c| \lesssim 10^{-2}$ for YBCO according to Eq. (5.59), which is not very broad. In the presence of a strong magnetic field such that the pair cyclotron radius $r_0 = (\phi_0/2\pi B)^{1/2}$ is smaller than $\xi = \xi_0\varepsilon^{-1/2}$, the coherence length perpendicular to the magnetic field, the length scale perpendicular to the field is determined by r_0 rather than ξ, and Eq. (5.59) is replaced by

$$|\varepsilon| \lesssim \left(\frac{k_B}{\Delta C}\frac{B}{\phi_0\xi_{0\parallel}}\right)^{2/3}. \tag{7.2}$$

Here $\xi_{0\parallel}$ is the coherence length in the direction along B. If $\xi_{0\parallel} = \xi_{0c} = 2.4$ Å and $B = 5\,\text{T}$ are entered in Eq. (7.2), the width of the critical region is 2 K and cannot be neglected. Furthermore, when the field B is sufficiently strong, the lowest energy solutions ($n = 0$ solutions of Eq. (5.28)) remain important as fluctuations. Then, the energy of fluctuations depends only on variation in the direction of $\boldsymbol{B}(\parallel \hat{z})$. In this sense, it must be emphasised, the system becomes one-dimensional. Since long range order is proven to be absent in a true one-dimensional system, it is doubtful if vortex lines in this case also undergo a phase transition to a lattice state with long range order, as shown in Section 5.4. In fact, data such as the magnetisation and the Peltier effect in addition to the electrical resistance indicate a continuous cross-over from the normal state. Calculations based on nonlinear fluctuation theory agree fairly well with the observed value on the high temperature side, relatively close to T_{cH} as shown in Fig. 7.14.

2. **Melting of the vortex lattice** What will happen at lower temperature? The electrical resistance of an ordinary superconductor may remain finite due to sliding of the vortex lattice in an ideal system without pinning centres. In reality, however, the superconducting current appears just below H_{c2} since the vortex lattice is fixed by the pinning centres. Now in HTSC interesting new features appear. First of all, the region in which the magnetic field is almost uniform is wide since λ is very large. Therefore it is appropriate to call the structure a vortex lattice as in superfluid ^{4}He, rather than a flux lattice in an ordinary type 2 superconductor. Moreover, ξ_{0c} is so small ($\sim$ a few Å) for $\boldsymbol{B} \parallel \hat{c}$ that the coupling between vortices created in the adjoining CuO_2-planes is suppressed by a factor of $(\xi_{0c}/\xi_a)^2$ and hence a vortex line is easily bent. There is even a possibility that vortices in different planes are independent in the Bi system. As a result, it is considered that the vortex lattice starts to melt and becomes a vortex liquid far to the low temperature side of $H_{c2}(T)$ as shown in the phase diagram of Fig. 7.15. (The lower critical field $H_{c1}(T)$ is almost degenerate with the line $H = 0$ since the GL parameter κ is large.) The vortex liquid cannot be distinguished from a state with strong fluctuations mentioned above and undergoes a continuous transition to the normal state. It is not clear whether there exists a definite melting curve in an ideal system or not. The vortex lattice may be stabilised by pinning centres if they exist. There is a possibility that a vortex glass state appears if the pinning centres are randomly distributed.

It should be added that the topic mentioned above is just one of the problems associated with HTSC in a magnetic field; there are other interesting issues such as the critical current.

Table 7.2. *The Sommerfeld constant and the magnetic susceptibility of heavy fermion systems.* $\gamma(0)/\chi(0)$ *is the ratio per cubic cm*

	$\gamma(0)$ $(mJ \cdot mol^{-1} \cdot [cm^{-3}] \cdot K^{-2})$	$\chi(0)$ $(10^{-3}\ emu \cdot cm^{-3})$	$\gamma(0)/\chi(0)$
$CeAl_3$	1620 [18.5]	0.41	45
$CeCu_2Si_2$	1000 [20]	0.13	153
URu_2Si_2	180 [3.66]	0.03/0.10	37
UCd_{11}	840 [5.21]	0.24	22
UPt_3	450 [10.6]	0.19	56
UBe_{13}	1100 [13.5]	0.18	75
{Na	1.5 [0.063]	0.001	63}

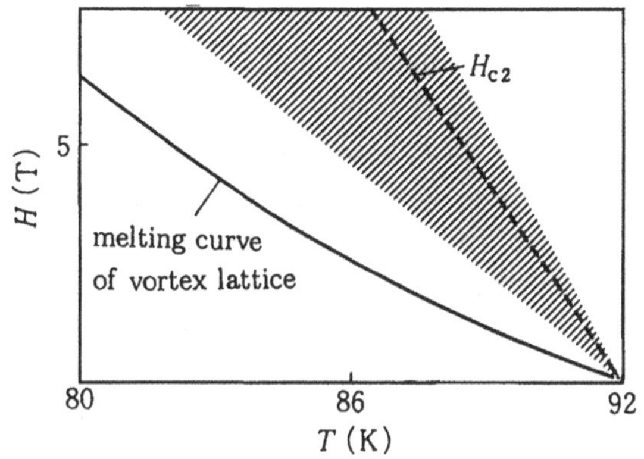

Fig. 7.15. The phase diagram of YBCO in a magnetic field. The shading indicates the cross-over area.

7.3 Superconductivity in heavy electron systems

There are so-called 'heavy fermion systems' among the compounds of Ce and U with f-electrons, which show extremely interesting properties. Table 7.2 lists some of them. Figure 7.16 shows the crystal structure of UPt_3 as an example. The physical properties of these materials at relatively high temperatures can be understood based on a system consisting of magnetic moments localised on Ce or U, of the same magnitude as in the atomic form, and of conduction electrons found in ordinary metals. However the physical properties change drastically below 200 $\sim$ 50 K, the exact value of which depends on the system under consideration. The number $\gamma(T) \equiv C(T)/T$, which corresponds to the Sommerfeld constant in an ordinary metal, and the susceptibility $\chi(T)$ are very large and the extrapolated values as $T \to 0$

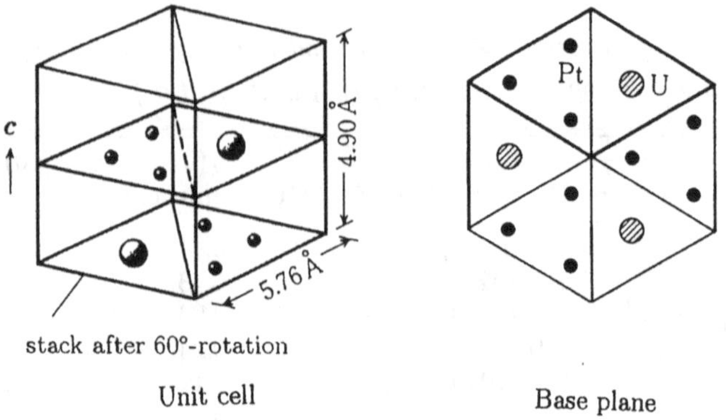

Fig. 7.16. The crystal structure of UPt_3.

are larger than those of an ordinary metal such as Na by a few orders of magnitude, as shown in Table 7.2. Note however that the ratio γ/χ is of the same order as that of Na. These observations lead to the conclusion that at low temperatures these systems have extremely large density of states $N(0)$ on the Fermi surface. In other words, they are described as liquids of dressed fermions whose effective mass m^* is $10^2 \sim 10^3$ times larger than the electron mass m. This has been confirmed by transport phenomena data, in particular measurement of the de Haas–van Alphen effect. Moreover, many of these systems undergo a phase transition to an antiferromagnetic state or a superconducting state at low temperatures. There are systems such as UPt_3, which become superconducting, that show strong antiferromagnetic (large q) as well as paramagnetic (small q) fluctuations, as observed by neutron scattering. It is characteristic of these materials that the amplitude of the magnetic moment in the antiferromagnetic state is reduced by $1/2 \sim 1/10$ from the atomic value.

The reason why they become heavy fermion systems at low temperatures and why antiferromagnetism with a small magnetic moment appears are discussed in [A-6] and will not be discussed here. Instead, we will be concerned here with the superconducting properties of heavy fermion systems. It should be added, however, that in a heavy fermion system the mixing of the s- and d-electrons with large band widths and the f-electrons is extremely small. Consequently the Coulomb repulsion between the f-electrons on the same site is large and hence the system is strongly correlated. In the latter sense, therefore, the system corresponds to a three-dimensional version of the model Hamiltonian (7.1) with the replacements d $\rightarrow$ f and p $\rightarrow$ s and with the condition $t_{\mathrm{fs}} \ll t_{\mathrm{ss}}$.

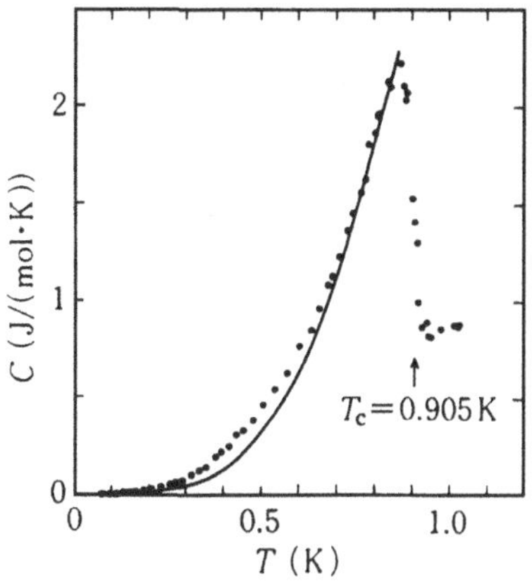

Fig. 7.17. The temperature dependence of the specific heat of UBe_{13}. The solid line is the theoretical curve assuming the ABM state of the ^{3}P pairing. (From [G-30].)

7.3.1 The superconducting state

It is important, first of all, to note that it is the heavy electrons that become superconducting since the specific heat jump ΔC at the transition temperature T_c is of the order of $C(T_c)$, see Fig. 7.17. The question is the type of pairing. The clue to this is the temperature dependence of various physical quantities. Such quantities as the specific heat C, the magnetic susceptibility χ, the ultrasound attenuation coefficient α, the NMR relaxation rate T_1^{-1} and so on decrease exponentially at low temperatures in an s-wave superconductor due to the energy gap as described in Chapter 3. In contrast with this, it is possible that the energy gap vanishes on points or along lines on the Fermi surface for a non-s-wave pairing as noted in Chapter 6 for the superfluid ^{3}He with a ^{3}P pairing. As shown in Table 7.3, the above quantities in such a state decrease in powers of T (T^2 for the points and T^3 for the lines) instead of exponential dependence on T. Figure 7.18 depicts the temperature dependence of T_1^{-1}, which does not show growth due to the s-wave coherence factor just below T_c, as does a copper oxide superconductor. In addition to these, the temperature dependences of $H_{c2}(T)$ and the London penetration depth $\lambda(T)$ are reported to deviate from those of the BCS theory. Thus it seems that a non-s-wave pairing is responsible for the superfluidity in many heavy fermion systems. Especially in the cases of UPt_3 and $U_{1-x}Th_xBe_{13}$ ($x \simeq 0.02 - 0.04$), which show more than two superconducting phases, this is beyond any doubt.

The temperature dependence of the specific heat of UPt_3 is shown in Fig. 7.19. One sees there are two peaks in the curve. The phase diagram (Fig. 7.20) in the $H - T$ plane has been obtained from this measurement

Table 7.3. *The antiferromagnetic transition temperature* T_N *(the Néel temperature) and the superconducting critical temperature* T_c. *The symbol* $\times$ *denotes that the corresponding phase transition has not been observed to date. The third column denotes the temperature dependence of the specific heat*

	T_N(K)	T_c(K)	$C(T < T_c)$
$CeCu_2Si_2$	(0.7?)	0.65	$T^{2.4}$
URu_2Si_2	17	1.5	T^2
UPt_3	5	0.5	T^2
UBe_{13}	?	0.9	$T^{2.9}$
UNi_2Al_3	4.6	1.0	
UPd_2Al_3	14	2.0	
UCd_{11}	5	$\times$	—
$CeAl_3$	$\times$	$\times$	—

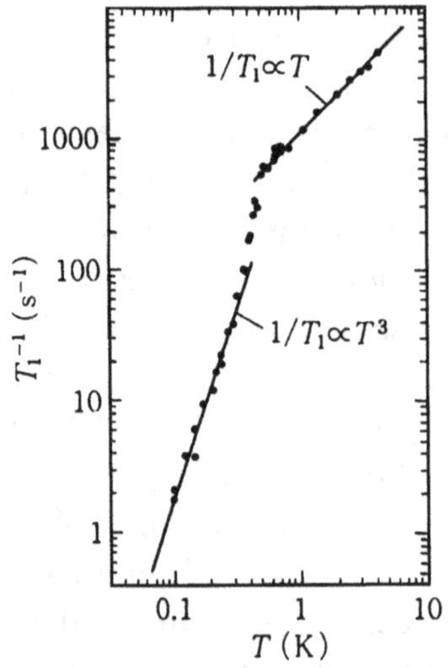

Fig. 7.18. The spin relaxation rate of ^{195}Pt in UPt_3. (From [G-31].)

as well as the ultrasound attenuation and the thermal expansion coefficient. According to this diagram, there are three superconducting phases A, B and C and four boundary lines sharing a single critical point. It seems that the superconducting phase of UPt_3 has a multi-component order parameter, i.e., has a non-s-wave pairing, since there are several superconducting phases. This is analogous to the A phase of superfluid ^{3}He, which splits into A

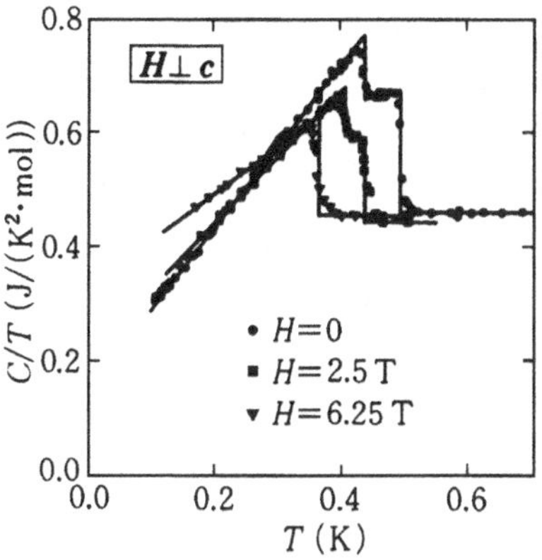

Fig. 7.19. The specific heat of UPt_3. Observe that there are two jumps. The jumps shift towards lower temperatures in the presence of a magnetic field. (From [G-32].)

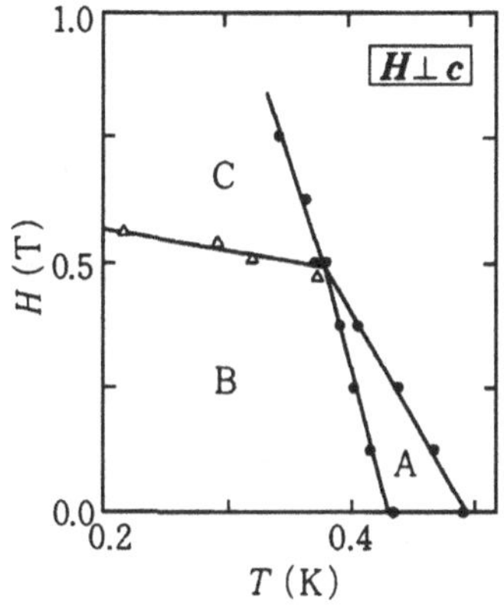

Fig. 7.20. The phase diagram of UPt_3 obtained from the specific heat, the upper critical field H_{c2} and the ultrasound attenuation. It is considered that there are three distinct superconducting phases A, B and C. (From [G-32].)

phase and A_1 phase in the presence of a magnetic field. It is estimated from H_{c2} values at low temperatures that $\xi_0 \sim 300\,\text{Å}$ and $\kappa \gg 1$ since λ_L, being proportional to $\sqrt{m^*}$, is large. Accordingly in the phase diagram the system is thought of as in a vortex lattice state in the region with a finite magnetic field.

What is the mechanism for superconductivity in these systems? At temperatures (energies) sufficiently lower than the Kondo temperature T_K, the system is regarded as a Fermi liquid made of heavy electrons and $k_B T_K$ is considered to play the rôle of the Fermi energy ε_F. The energy $k_B T_K$ is smaller than the Debye phonon energy ω_D in many heavy electron systems. Therefore the number of phonons contributing to the attraction between electrons is limited and, moreover, the retardation effect cannot be expected, which makes superconductivity by the phonon mechanism less probable. Rather it is expected that spin fluctuations play the central rôle since electrons in the system are strongly correlated. Note, however, that even the theory of a heavy electron system in the normal state is still incomplete and hence the quantitatively correct microscopic theory of the superconducting state has not yet been obtained. Therefore semi-phenomenological considerations using the group theoretical classification are now exclusively employed; the elements of this will be briefly surveyed next.

7.3.2 Symmetry of pairing in crystal

Pairing has been considered in Chapter 2 with assumptions that the system is invariant under an arbitrary space rotation and, in particular, that the interaction between particles is written in the form of Eq. (2.12), namely

$$V(\boldsymbol{k},\boldsymbol{k}') = \sum_l (2l+1) V_l P_l(\hat{\boldsymbol{k}},\hat{\boldsymbol{k}}'), \tag{7.3}$$

where V_l is regarded as a constant for $k, k' \sim k_{\rm F}$. The gap equation in the BCS weak-coupling theory in this case takes the form (2.58) for both a spin-singlet ($S = 0$) pairing and a spin-triplet ($S = 1$) pairing. Moreover, the equation which determines $T_{\rm c}$ is obtained from Eq. (2.60) as

$$\Delta_{\alpha\beta}(\hat{\boldsymbol{k}}) = -v_{\rm c}{}^{-1} \int {\rm d}\Omega_{k'} V(\hat{\boldsymbol{k}},\hat{\boldsymbol{k}}')\Delta_{\alpha\beta}(\hat{\boldsymbol{k}}') \tag{7.4}$$

since one may put $\varepsilon_{k\alpha} = |\xi_{k\alpha}|$. Here $v_{\rm c}$ is given by

$$v_{\rm c}{}^{-1} = \ln(1.13\ \omega_{\rm c}\beta_{\rm c})$$

in the weak-coupling theory and the integration yields the angular average over the Fermi surface. This is therefore a linear eigenvalue equation to determine $\beta_{\rm c} \equiv (k_{\rm B}T_{\rm c})^{-1}$.

Determining the kind of superconducting states obtained as a solution to this equation is the same problem as the classification of wave functions according to the symmetry of the Hamiltonian and can be dealt with by obtaining irreducible representations of the symmetry group. It is well-known that there is a $2 \times (2l+1)$-dimensional irreducible representation for a spin singlet ($S = 0$) pair and a $2 \times 3 \times (2l+1)$-dimensional one for a spin triplet ($S = 1$) pair for each l if the system is isotropic. For instance, the ^{3}P-pair can be represented by nine complex quantities $A_{\mu j}$. All the states belonging to the same irreducible representation have the same $T_{\rm c}$ while those with different l

have different T_c. In other words, l (and S) specifies a degenerate class. There are thus infinitely many possible classes since the only requirement is an attractive V_l.

Then what kind of superconducting state is realised in a given crystal and how many classes of degenerate states are possible? It must be assumed, when considering the pairing in a crystal, that the density of states of the pairing electrons, the interaction between electrons and so on have only the same symmetries as those of the crystal. Let us restrict ourselves to the case of a uniform superconducting state without an external field. Suppose for simplicity that the spin–orbit interaction is so strong that the invariance under separate rotations of the orbit space and the spin space of the pairs is frozen. (It should be noted, however, that spin–orbit interaction in UPt_3 is weak according to recent research and hence the relative rotation between the spin and orbit spaces must be considered.) Then the symmetry under question in the 'symmetry breaking' due to pairing is the product of the rotational symmetry G_p of the crystal, the gauge transformation and the time reversal,

$$G_p \times U(1) \times T. \tag{7.5}$$

An example of time-reversal symmetry breaking is the case with non-unitary pairing ($\Delta^* \neq \Delta$), see Section 2.3.1. Since the order of the point group G_p is finite, there are only a finite number of possible classes in contrast with an isotropic case (the degrees of freedom such as those represented by $f(\boldsymbol{k})$ in Eqs. (7.6) and (7.7) below are not considered). Note also that such expressions as the s-wave pair and the p-wave pair in a crystal do not have the same meaning as in an isotropic case. Leaving detailed discussions to [E-7], [E-8] and [E-9], let us consider a crystal with a simple cubic symmetry by way of example. In the case where G_p is the point group of a simple cubic system, there are five irreducible representations both for $S = 0$ and $S = 1$ and therefore there are five classes of states with different T_c. These are two 1-dimensional representations, one 2-dimensional one and two 3-dimensional ones. Here we list the bases of the 1-dimensional representations only:

$$S = 0 \quad \Delta(\boldsymbol{k}) = f(\boldsymbol{k}) \tag{7.6a}$$

$$S = 1 \quad \boldsymbol{\Delta}(\boldsymbol{k}) = (\hat{\boldsymbol{x}}k_x + \hat{\boldsymbol{y}}k_y + \hat{\boldsymbol{z}}k_z)f(\boldsymbol{k}) \tag{7.6b}$$

$$S = 0 \quad \Delta(\boldsymbol{k}) = (k_x{}^2 - k_y{}^2)(k_y{}^2 - k_z{}^2)(k_z{}^2 - k_x{}^2)f(\boldsymbol{k}) \tag{7.7a}$$

$$S = 1 \quad \boldsymbol{\Delta}(\boldsymbol{k}) = \{\hat{\boldsymbol{x}}k_x(k_y{}^2 - k_z{}^2) + \hat{\boldsymbol{y}}k_y(k_z{}^2 - k_x{}^2) + \hat{\boldsymbol{z}}k_z(k_x{}^2 - k_y{}^2)\}f(\boldsymbol{k}), \tag{7.7b}$$

where $\boldsymbol{\Delta}(\boldsymbol{k})$ is the d-vector defined by Eq. (2.40) and $f(\boldsymbol{k})$ is a function with a cubic symmetry. The base (7.6a) corresponds to s-wave superconductivity while Eq. (7.6b) is similar to the order parameter of the BW state (6.15) of the superfluid ^{3}He. The gaps of Eq. (7.6) are finite everywhere on the Fermi surface while those of Eq. (7.7a) and Eq. (7.7b) vanish along lines and on points on the Fermi surface respectively. This is due to the loss of independence of the rotations and the gauge transformations as was shown in the ABM state of the superfluid ^{3}He. It should be noted that the gap may vanish on points or along lines in a superconductor without internal degrees of freedom (where the only degree of freedom is the overall phase).

Let us give another example. Suppose we require in a hexagonal lattice system such as UPt_3 that the gap should vanish along a line so that the specific heat at $T < T_c$ is proportional to T^2 and, also, that there are more than two phases. Then the two-dimensional representation with a two-component order parameter given by

$$\Delta(\boldsymbol{k}) = \eta_1 k_z k_x + \eta_2 k_z k_y \tag{7.8a}$$

is the simplest form if we assume spin singlet ($S = 0$) pairing. (The function corresponding to $f(\boldsymbol{k})$ above is set equal to unity.) The coefficients η_1 and η_2 represent the two-component degrees of freedom. If, on the other hand, a spin-triplet pairing is assumed, it takes the form called E_{2u} given by

$$\boldsymbol{\Delta}(\boldsymbol{k}) = \hat{\boldsymbol{z}} k_z \left\{ \eta_1 (k_x^2 - k_y^2) + \eta_2 2 k_x k_y \right\}. \tag{7.8b}$$

Recently, in [G-33], the Knight shift of ^{195}Pt in UPt_3 has been measured down to 28 K and it was found that it does not change even in a superconducting state. If this observation is confirmed, the pairing in UPt_3 would be of spin-triplet form with odd parity. Whether this pairing is unitary or not is also discussed.

The gap equation becomes nonlinear at $T < T_c$ and the states which are degenerate at T_c no longer have the same free energy. The BW state or the ABM state in ^{3}He is a class of states as discussed in Chapter 6 and all the states belonging to the same class have the same free energy. The class forms a representation of a subgroup of the original group of symmetries. Ordinarily, using the free energy of the GL theory which retains the Δ^4 or Δ^6 terms, we try to find, with the help of group theory, possible classes of the order parameter and the associated energy gap in various crystals.

In a system with co-existing ferromagnetism or antiferromagnetism, one has to take into account its order parameter and the coupling with the superconducting order parameter. The coupling between the altered crystal structure and the anisotropic superconductivity is possibly an interesting question for the future.

Appendix 1
Bose–Einstein condensation in polarised alkaline atoms

A1.1 Condensate in confining potential

This subject is directly related to Chapter 1 of this book and will be briefly summarised here although it has already been treated in an appendix of [E-11].

Spin polarised hydrogen H$\downarrow$ is listed in the right hand column of Table 1.1 as a system predicted to show Bose-type superfluidity. Bose–Einstein condensation was observed, somewhat unexpectedly, in spin polarised alkaline gases Li$\downarrow$, Na$\downarrow$ and Rb$\downarrow$ with quite an ingenious method using laser light, prior to observation in H$\downarrow$ gas (see [H-1]– [H-3]). It is unquestionable that the macroscopic wave function appears although superfluidity has yet to be observed. See [H-4] for works on H$\downarrow$ and related systems preceding this breakthrough.

In a typical experiment the magneto-optical trap, which we can approximate by an anisotropic 3-dimensional harmonic oscillator potential, is used to provide the confining potential for atoms. Its spatial scale is given by $R \sim (\hbar/m\omega)^{1/2} \sim 10^{-3}$ cm, where m is the mass of the atom while ω is the frequency of the harmonic oscillator. There are $N \sim 10^6$ atoms trapped in a volume $\sim R^3$, for which the Bose–Einstein condensation temperature T_{BE} given by Eq. (1.3) is of the order of 10^{-7} K for Rb atoms. It should be added that evaporation cooling, which has been used to cool H$\downarrow$ gas, is required in addition to laser cooling to reach this temperature.

The condensate is described by the ground state wave function ψ_0 in the trapping well provided that interatomic interaction is neglected. It takes the form $\psi_0 \propto \exp\{-m\omega r^2/2\hbar\}$ if the potential is of the harmonic oscillator type, where the anisotropy is absorbed in rescaling. The atomic density at the centre is $n_{\mathrm{c}} \sim |\psi_0(0)|^2$, which is approximately $5 \times 10^{14}\,\mathrm{cm}^{-3}$ in an actual experiment.

In reality, there is an interatomic interaction. A repulsive force acts between two polarised (i.e., spin-aligned) ^{87}Rb atoms, for example, and the scattering length for a low energy spin-triplet s-wave is known to be $a \sim 100a_0$, where a_0 is the Bohr radius. If the healing distance ξ is evaluated according to Eq. (1.9) with density n_c at the centre, one obtains $\xi \sim 10^{-3}$ cm. Therefore R may or may not be larger than ξ depending on individual cases.

Let $V(\boldsymbol{x})$ be the confining potential energy. The equation that determines the condensate wave function $\psi_s(\boldsymbol{x})$, corresponding to Eq. (1.8), is

$$-\frac{1}{2}\xi^2\nabla^2\psi_s - \left(1 - \frac{m\xi^2}{\hbar^2}V(\boldsymbol{x})\right)\psi_s + |\psi_s|^2\psi_s = 0. \qquad \text{(A1.1)}$$

The first term is negligible if ξ is sufficiently smaller than R, for which case the density of the condensate $n_s(\boldsymbol{x}) = |\psi_s(\boldsymbol{x})|^2$ is given by

$$n_s(\boldsymbol{x}) = n_c\left[1 - (m\xi^2/\hbar^2)V(\boldsymbol{x})\right]. \qquad \text{(A1.2)}$$

Therefore the density profile of atoms is given by the inverted potential at low temperatures ($T \ll T_{BE}$). The spatial variation of the density is observed experimentally by measuring the absorption of an appropriate laser light.

A1.2 Dynamics of dilute BE condensate

Once the BE condensation of a gas of polarised alkaline metal is observed, it is quite natural that the dynamic properties of the gas should be studied, see [H-5] and [H-6]. It is the collective motion of the condensate at low temperatures $T \ll T_{BE}$ that should be considered first. Since the effect of thermal excitations is negligible in this case, the foundation of the theoretical arguments is the nonlinear Schrödinger equation (A1.1) with the term $\mathrm{i}(m\xi^2/\hbar)\partial\psi/\partial t$ added, that is,

$$\mathrm{i}\frac{m\xi^2}{\hbar}\frac{\partial\psi}{\partial t} = -\frac{1}{2}\xi^2\nabla^2\psi - \left(1 - \frac{m\xi^2}{\hbar^2}V\right)\psi + |\psi|^2\psi, \qquad \text{(A1.3)}$$

which is called the Gross–Pitaevskii equation in superfluid helium research. Suppose there is a condensate represented by the solution $\psi_s(\boldsymbol{x})$ of Eq. (A1.1). To study the motion of a small deviation from the condensate solution, that is, a low energy excitation, let us put the wave function of the deviation as

$$\psi - \psi_s = \phi\mathrm{e}^{-\mathrm{i}\omega t} \qquad \text{(A1.4)}$$

where the condition $|\phi| \ll |\psi_s|$ is assumed. Then only the first order terms in ϕ need be kept in Eq. (A1.3) and one obtains

$$\frac{m\xi^2}{\hbar}\omega\phi = -\frac{1}{2}\xi^2\nabla^2\phi - \left(1 - \frac{m\xi^2}{\hbar^2}V\right)\phi + 2n_0\phi + n_0\phi^* \tag{A1.5}$$

and a similar equation for ϕ^*. Here $\psi_s = \sqrt{n_0(\boldsymbol{x})}$ is assumed to be real for simplicity.

Let us next consider what excitations are obtained for a uniform dilute Bose gas (i.e., $V(\boldsymbol{x}) = 0$ and $n_0 = \bar{n}_s$). Since the system is uniform, the wave function ϕ is a plane wave $\phi \propto e^{i\boldsymbol{p}\cdot\boldsymbol{x}/\hbar}$. The corresponding excitation spectrum, found from Eq. (A1.5) is

$$\varepsilon_{\boldsymbol{p}}^2 = \varepsilon_{\boldsymbol{p}}^0(\varepsilon_{\boldsymbol{p}}^0 + 2\hbar^2/m\xi^2), \tag{A1.6}$$

where $\varepsilon_{\boldsymbol{p}}^0 = p^2/2m$. It turns out from Eq. (A1.4) that this excitation is a density fluctuation, namely a phonon, and from Eq. (A1.6) that the energy approaches cp in the long wavelength limit ($p \to 0$). Here the sound velocity c is given by $c = \hbar/m\xi = \sqrt{\bar{n}_s g/m}$. This fact is essential in superfluidity, see Section 1.4 where it has been shown that the elementary excitation energy in the rest frame is given by $\varepsilon_{\boldsymbol{p}} + \boldsymbol{p}\cdot\boldsymbol{V}_s$ when the whole system moves with velocity $\boldsymbol{V}_s$. Here $\boldsymbol{p}$ and $\varepsilon_{\boldsymbol{p}}$ are quantities in the frame moving with velocity $\boldsymbol{V}_s$. If the spectrum is of the phonon type $\varepsilon_{\boldsymbol{p}} = cp$, as in the present case, the system energy cannot be lowered by creating an elementary excitation when $|\boldsymbol{V}_s| < c$. In other words, the flow does not decay. We should note that interaction must be present for the phonon velocity to be finite. It should also be added that the Bogoliubov transformation defined in Section 2.4 was first introduced to obtain the spectrum (A1.6) from Eq. (A1.5) in a dilute Bose gas model.

The situation becomes far more complicated in the case of a condensate in a magneto-optical trap. Lower frequency collective excitations can also appear as deformations of the condensate. Let us estimate the order of the frequency ω_s when $R \gg \xi$ is satisfied. Note first that ω_s is given by the inverse of the time with which a phonon propagates through a system of size R. Therefore it follows that $\omega_s^{-1} \sim R/c$. Suppose the trapping potential is of the harmonic oscillator type with $V \sim \frac{m}{2}\omega_0^2 r^2$ and the radius R is defined by the radius for which $n(r) = 0$ in Eq. (A1.2). Then one finds $R \sim c/\omega_0$ and, as a result, $\omega_s \sim \omega_0$. Note that this is independent of the density of the condensate. In fact, it has been shown that, when the potential is of the harmonic oscillator type, there is a mode with frequency $2\omega_0$, which corresponds to oscillation of the condensate without change of shape.

Experimentally, a collective motion may be excited by applying a magnetic field with an appropriate frequency. The results obtained so far seem to roughly agree with theoretical results based on the nonlinear Schrödinger equation, see [H-7].

Even when hydrodynamic collective motion is observed with no BE condensate, it has been confirmed that the frequency shifts as the system undergoes condensation. It is also observed that the lifetime of a collective motion becomes much longer in the presence of a condensate.

Superfluidity in the present system has not yet been observed. To confirm superfluidity explicitly, one must check whether or not the circulation around a vortex line, generated by rotation, is quantised in units of κ defined by Eq. (1.2). For example, it is expected that a single quantised vortex line appears in the condensate at angular frequency Ω_{c1}, which corresponds to H_{c1} in a type 2 superconductor, and one can then observe the change in density distribution of the condensate associated with it. We also expect that there is a phenomenon similar to the Josephson effect when two condensates, separated spatially, are weakly coupled.

To close this brief review we should note that attempts are being made to extract a coherent atom pulse, that is, a pulsed atom laser from the trapped condensate (see [H-8] for example).

Appendix 2

Recent developments in research on high temperature superconductors

A2.1 Pair structure of high temperature superconductors and Josephson effects

It is current opinion that the pair in a copper oxide superconductor is not spin-singlet s-wave type but has the symmetry of a spin-singlet d-wave pair.

It is considered, from the temperature dependence of the specific heat, the penetration depth, nuclear magnetic relaxation rate and so on, that the energy gap vanishes on points or lines on the Fermi surface (see Section 7.2). Recent experiments using angular resolution photoelectron emission revealed that the gap is vanishingly small in the diagonal directions, $|k_x| = |k_y|$. These facts may be explained if the order parameter $\Delta(\boldsymbol{k})$ in momentum space takes the d-wave form

$$\Delta(\boldsymbol{k}) = (\cos k_x - \cos k_y)\Delta \tag{A2.1}$$

with the k-dependence shown in Fig. A2.1(a). Here the system is regarded as two-dimensional and the x- and y-axes are taken as in Fig. 7.7. Let us recall that Δ is a complex number and the spatial variation of its phase χ produces the supercurrent. The above experiments can also be accounted for by s-wave pairing with quite a large anisotropy as shown in Fig. A2.1(b). Theorists are of the opinion that antiferromagnetic fluctuations, which result in d-wave pairing, are important in copper oxides, although there is no direct evidence for the claim so far. From 1993 onwards, there have been several experiments making use of the Josephson effect to determine the structure of the pairing. These experiments have produced interesting results that will be outlined below.

If the pairing takes the form shown in Fig. A2.1(a), then the order parameter changes sign, in other words, the phase changes by π depending whether $\boldsymbol{k}$ is directed towards the a-axis or the b-axis in the CuO_2 plane. One may

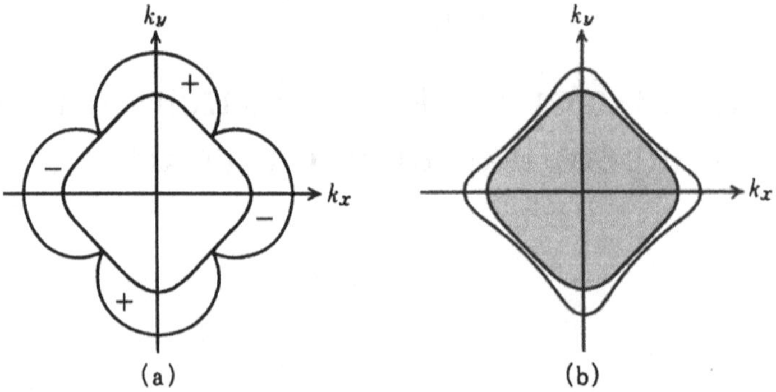

Fig. A2.1. (a) The amplitude and signature of the d-wave pairing on the Fermi surface. (b) The amplitude of the anisotropic s-wave pairing.

conclude that the pairing is not s-wave but d-wave if this phase difference is detected experimentally. The simplest experiment employs the tunnelling device shown in Fig. A2.2, see [H-9]. Here a loop is made of the tunnelling junctions of an s-wave superconductor such as Pb attached to the faces a and b of a single crystal of a high-T_c superconductor (YBCO in this experiment). We expect that electron pairs whose momenta $(\boldsymbol{k}, -\boldsymbol{k})$ are perpendicular to the face of the junction mainly tunnel through (although it seems that for the present this claim is not clearly founded on microscopic theory). If this is the case, the phase of the pair tunnelling through the side a into YBCO is different by π from the phase of that through b. In other words, the phase of a pair when it goes around the loop changes by π. As a consequence, the loop must carry current. In the limit of a large self-inductance, this current produces a stable state with a flux $\phi_0/2$ ($\phi_0 = hc/2e$) even without an external magnetic field. One would like to detect this flux but in practice this is not easy, due to problems associated with residual magnetic flux or flux due to the measurement current. Accordingly the method employing the interference effect, introduced in Section 3.4, is used, see [H-10].

One may assume that variation in the phase χ takes place only at the two tunnelling junctions in the loop of Fig. A2.2 as assumed in Section 3.4. Then it holds that

$$\chi_R(1) - \chi_L(1) + \chi_L(2) - \chi_R(2) + \delta\pi = 2\pi\phi, \tag{A2.2}$$

where ϕ is the flux through the loop in the unit of ϕ_0. The parameter δ is ± 1 if the phase change π (π-shift) is associated with YBCO in the right side (R) of the junction, while $\delta = 0$ if the order parameter of R is of s-wave type. Let us assume for simplicity that the two junctions have the same

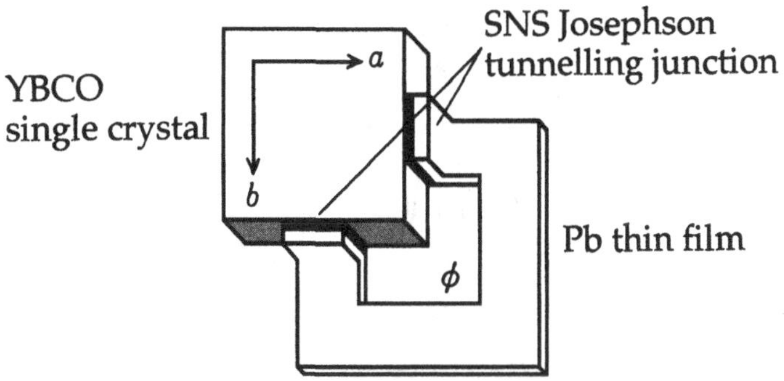

Fig. A2.2. A loop with π-shift. (From [H-9].)

characteristics. The bias current I from left to right in this case is given by

$$I/I_c = \sin(\chi_0 + \pi\phi) + \sin(\chi_0 - \pi\phi + \delta\pi), \tag{A2.3}$$

where I_c is the critical current of each junction and

$$\chi_0 \equiv \chi_R(1) - \chi_L(1) - \pi\phi.$$

If $\delta = 0$, one reproduces the result

$$I/I_c = 2\sin\chi_0 \cos\pi\phi \tag{A2.4}$$

obtained in Section 3.4, while if $\delta = \pm 1$ one has

$$I/I_c = 2\cos\chi_0 \sin\pi\phi. \tag{A2.5}$$

Figure 3.10 shows I/I_c as a function of the flux ϕ for $\delta = 0$, from which we see maximal values occur when ϕ is an integer. In contrast, it takes a maximal value when ϕ is a half-odd integer if $\delta = \pm 1$. Thus the difference shown in Fig. A2.3 must occur. In other possible experiments we measure with the help of another SQUID the magnetic flux produced by the current circulating in the loop when the bias current is kept at zero; see [H-11]. We will not go into details of the experimental results here. Although most results favour d-wave pairing, there is a report [H-12] in which the π-shift has not been observed.

The above Josephson circuit is symmetric with respect to time-reversal transformation. A state with the bias current I and the magnetic flux $\delta\phi_0$ is mapped to a state with reversed current $-I$ and reversed flux $-\delta\phi_0$ under time reversal. If a circuit is made of an ordinary superconductor, a state with $\delta = 0$ is mapped to itself, while a state with the π-shift is necessarily mapped to a different state. For example, a state with $n = 1$ is mapped to

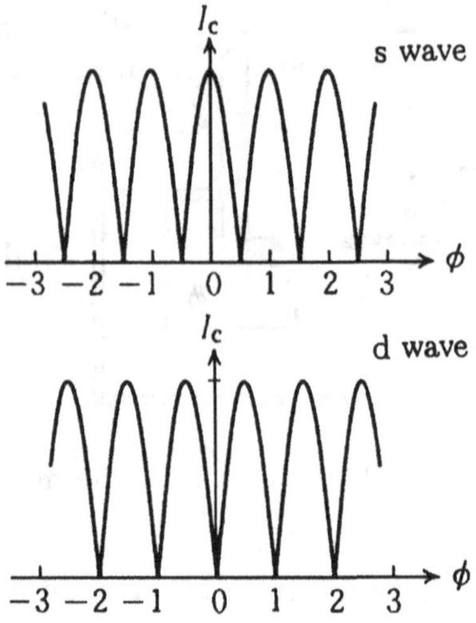

Fig. A2.3. The critical current as a function of the flux through the loop.

that with $n = -1$, see [H-13]. Note also that superconductivity with such a complex order parameter as

$$\Delta(\boldsymbol{k}) = (k_x^2 - k_y^2)\Delta_1 + \mathrm{i}k_x k_y \Delta_2$$

or

$$\Delta(\boldsymbol{k}) = (k_x^2 - k_y^2)\Delta_1 + \mathrm{i}\Delta_s$$

does not have this symmetry (unless $\Delta_2 \to -\Delta_2$ and $\Delta_s \to -\Delta_s$ under time reversal). This observation may be one of the clues to identify the symmetry of the pair.

So far we have discussed the spatial structure of the order parameter, that is, the $\boldsymbol{k}$-dependence of the pair wave function. On the other hand, the ω-dependence of Δ as mentioned in Chapter 4 is also interesting in HTSC. However the experiment seems to be difficult in HTSC since the coherence length is extremely small.

A2.2 High temperature superconductor in a magnetic field

The characteristics of a high temperature superconductor in a magnetic field have been studied in Section 7.2, where it was pointed out, in conjunction with its peculiar layered structure, that fluctuations are quite important and that there is a possibility that an ordinary Abrikosov lattice melts to form a vortex liquid state. This unique physics, realised for the first time due to the discovery of HTSC, has been actively studied since then and interesting results obtained. Although this can be observed in YBCO, most studies have used a typical HTSC, BSCCO ($Bi_2Sr_2CaCu_2O_8$). The latter has a strong anisotropy and the effect of fluctuations is far more important.

An ordinary type 2 superconductor undergoes a second order phase transition at a temperature T determined by $H = H_{c2}(T)$ in the presence of an external field H and is in the Abrikosov vortex lattice state below that temperature. In contrast with this, there does not exist such a phase transition in HTSC when a magnetic field is applied parallel to the c-axis. Rather it undergoes continuous transition from the normal state to the vortex liquid state. A vortex lattice state should appear, however, as the temperature is further lowered. Then it is conjectured that, in an ideal system without such effects as a pinning, there exists a clear-cut thermodynamic phase transition between the vortex liquid and the lattice. It is also likely that the transition is of the first order, like an ordinary liquid–gas transition, since the transition from the normal state is pushed down to low temperatures by the effect of fluctuations. These predictions have been based mainly on the temperature dependence of the electrical resistivity in a magnetic field. Recently they were confirmed directly by thermodynamic measurements and quite interesting facts have come to light.

In [H-14], an external field $\boldsymbol{H}(\parallel c)$ was applied to a planar BSCCO crystal ($T_c = 90\,\mathrm{K}$) perpendicular to the c-axis and the magnetic flux density B at the sample surface was measured with precision making use of the Hall effect of the two-dimensional electron system in GaAs. Figure A2.4 shows the sharp jump in the flux density B observed as the temperature is lowered, while Fig. A2.5 is the phase diagram in the temperature–magnetic field plane. This is the first order phase transition, since B has a jump, and B is larger on the liquid side where the vortex density is larger, as in the water–ice transition. The jump in Fig. A2.4 corresponds to an entropy jump $\Delta S = 0.3k_B$ per vortex line whose length is the interlayer distance. Note that the melting curve $B_m(T)$ is far below $H_{c2}(T)$ in BSCCO. What is more interesting is that this melting curve looks as if it ends at the critical point ($T = 37\,\mathrm{K}$ and $B = 380\,\mathrm{G}$). In other words, the jump in B disappears there. It may be possible that the phase transition is of the second order beyond this point due to the effect of randomly distributed defects.

How is the phase transition between the vortex liquid and the vortex lattice reflected in the electrical resistance? Let us note that a similar first order phase transition is also observed in a clean YBCO. Figure A2.6 shows the temperature dependence of the electrical resistance along with the jump in magnetisation, see [H-15].

It can be seen from this that the resistivity suddenly disappears as the

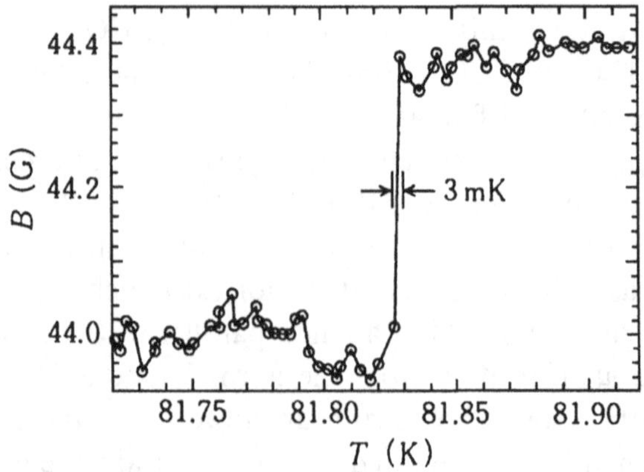

Fig. A2.4. A jump in the magnetic flux density B observed as the temperature is lowered with an external magnetic field of 53 Oe (note that 1 G = 10^{-4} T). (From [H-14].)

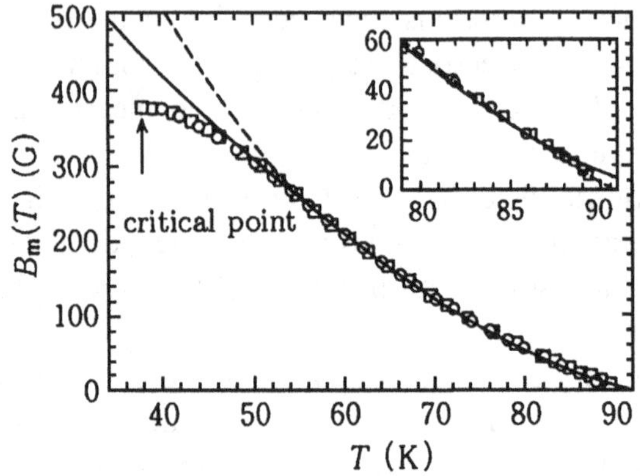

Fig. A2.5. Phase diagram showing the phase boundary between the vortex lattice and the vortex liquid in BSCCO. (From [H-14].)

vortex liquid solidifies. It is considered as the origin of this behaviour that the pinning mechanism starts to work as the vortices form a lattice with stiffness and the decay of current due to vortex motion is then forbidden. It is also shown in Fig. A2.7 that the changes of H and T and the measurement of ρ all give the same melting curve. The jump disappears if the sample is irradiated by an electron beam to create defects, after which $\rho(T)$ approaches zero as T is lowered. It seems that the pinning mechanism does not work efficiently in BSCCO and the discontinuous change in ρ has not been observed.

Another possible origin of the phase transition, which is not considered above, is that the links of the vortex lines in different layers may collectively

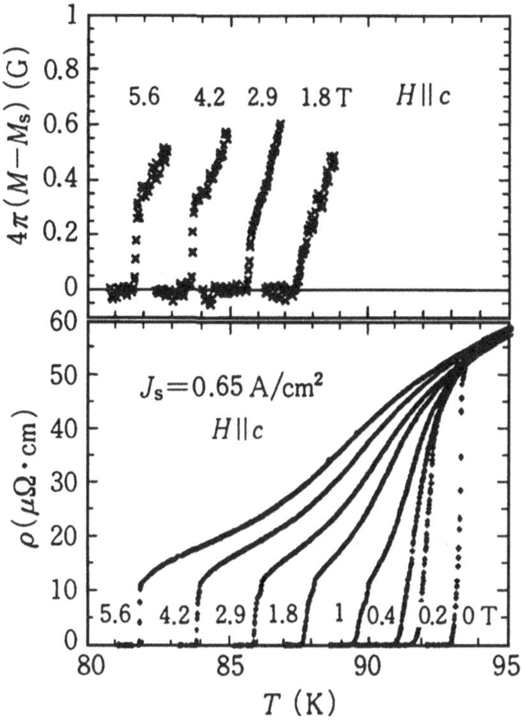

Fig. A2.6. The jump in magnetisation and disappearance of the electrical resisitivity ρ in YBCO. The numbers in the figure denote the magnitude of the external field ($H \simeq M$). (From [H-15].)

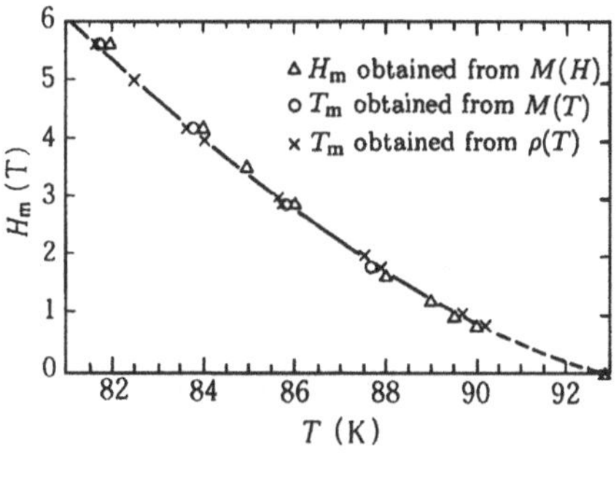

Fig. A2.7. The melting curve of the vortex lattice in YBCO obtained from magnetisation and resistivity measurements.

disappear. It may also be expected that the vortex glass state appears in a disordered crystal. So far the theories related to these problems are either phenomenological or based on simulations. It is interesting to analyse theoretically how the phase correlation between order parameters at two points

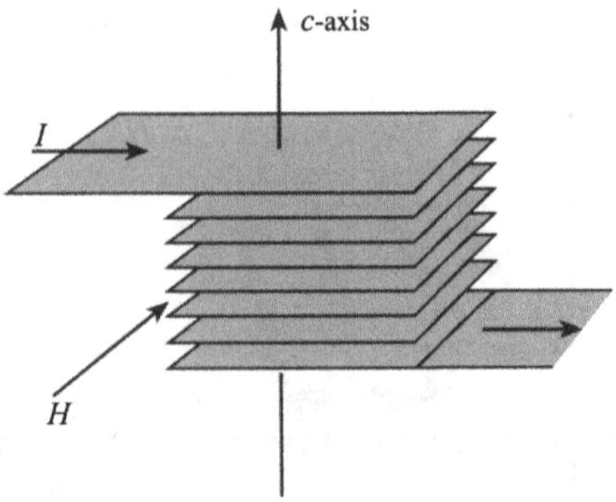

Fig. A2.8. CuO_2 sheets coupled through Josephson junctions.

separated by R and the correlation of the positions of the vortex centres change as R increases. The former is the very parameter that characterises superconductivity and superfluidity. These correlations should, of course, behave differently in a clean system and in a disordered system. It seems that the entire picture of the physics of the vortex state in HTSC is yet to be clarified.

A2.3 Supplement to research in high-T_c superconductors

1. It was reported in Chapter 7 that the superconductor with the highest T_c was $Bi_2Sr_2CaCu_2O_8$ with $T_c = 125\,K$. Shortly after this discovery, it was found that a group of materials with similar structure but incorporating mercury becomes superconducting (1992). Among others, $HgBa_2Ca_2Cu_3O_{8-x}$ is found to have $T_c \simeq 135\,K$ ($\sim 150\,K$ under pressure). This record is not broken at the moment (see [H-16]). Unfortunately this material is unstable and a good sample cannot be prepared. Thus its properties are seldom studied.
2. It was found recently that Sr_2RuO_4, which has a layered perovskite structure like La_2CuO_4 with Cu substituted by Ru in the platinum group, becomes superconducting below $T_c \simeq 1\,K$, see [H-17]. This material has a structure shown in Fig. 7.6(a) with La substituted by Sr and Cu by Ru and shows strong two-dimensionality even without doping (see [H-18]), in contrast with La_2CuO_4, which becomes conducting only when La is substituted by Sr by doping. The low $T_c \simeq 1\,K$ suggests that what is important for high-T_c superconductivity is not simply a layered structure but the existence of the CuO_2 sheets.

3. HTSC has a strong two-dimensionality due to the layered structure. It has been considered in particular that the superconducting properties may be explained by regarding the system as weakly coupled CuO_2 sheets. If this is the case, it may be possible to observe the Josephson effect due to the microscopic structure itself. Recently, as microfabrication technology has advanced, a sample with a few tens of microns edge size and less than 100 CuO_2 sheets has been made and the d.c. and a.c. Josephson effects were indeed observed. Figure A2.8 shows the schematic diagram of the system. The current $I = I_c(H)$ is measured in the d.c. mode. This effect is also observed in both BSCCO and YBCO with weaker interlayer coupling made by reducing oxygen atoms, see [H-18].

References and bibliography

Books on many-body problem and related subjects

[A-1] A.A. Abrikosov, L.E. Dyaloshinskii and L.P. Gor'kov: *Methods of Quantum Field Theory in Statistical Physics* (Dover, New York 1963).
A typical textbook on the many-body problem based on the Green's functions. Fermi liquids, superfluidity of Bose systems and superconductivity are treated in detail.

[A-2] A.A. Abrikosov: *Fundamentals of the Theory of Metals* (North-Holland, Amsterdam 1988).
A thick book with 630 pages. The first half (Part I) is devoted to the Fermi liquid theory of metals while the second half (Part II) covers the theory of superconductivity.

[A-3] P.W. Anderson: *Basic Notions of Condensed Matter Physics* (Benjamin, Menlo Park 1983).
Unique expositions and a collection of original papers on symmetry breaking, Fermi liquid theory and other subjects.

[A-4] N.W. Ashcroft and N.D. Mermin: *Solid State Physics* (Holt, Rinehart and Winston, Saunders College, Philadelphia 1976).
Chapter 35 provides an introduction to superconductivity.

[A-5] A. Fetter and J.D. Walecka: *Quantum Theory of Many-Particle Systems* (McGraw-Hill, New York 1971).
A textbook similar to but easier to understand than [A-1].

[A-6] K. Yamada: *Electron Correlation in Metals* (Cambridge University Press, Cambridge, to be published).

Books on superfluidity

[B-1] K.H. Benneman and J.B. Ketterson: *The Physics of Liquid and Solid Helium, Parts I and II* (Wiley-Interscience, New York 1976).

A collection of excellent expositions on superfluid ^{4}He, normal and superfluid ^{3}He.

[B-2] R.J. Donnelly: *Experimental Superfluidity* (University of Chicago Press, Chicago 1967).

The most easily understood book on superfluid ^{4}He.

[B-3] R.J Donnelly: *Quantized Vortices in Helium II* (Cambridge University Press, Cambridge 1991).

An excellent book on the physics of superfluid ^{4}He, which is only briefly treated in the present book.

[B-4] W.E. Keller: *Helium-3 and Helium-4* (Plenum Press, New York 1969).

A book covering the general aspects of helium systems, although slightly out of date.

[B-5] I.M. Khalatnikov: *Introduction to the Theory of Superfluidity* (Benjamin, New York 1965).

The theory of superfluids with emphasis on hydrodynamic aspects.

[B-6] F. London: *Superfluids* (John Wiley, New York Vol. I 1950, Vol. II 1954).

Vol. I is on superconductivity while Vol. II is on superfluidity of liquid ^{4}He. These classics are worth reading, even today.

[B-7] D. Vollhardt and P. Wölfe: *The Superfluid Phases of Helium 3* (Taylor and Francis, London 1990).

An exposition of more than 600 pages exclusively on the research on superfluid ^{3}He as of the end of the 1980s.

[B-8] H. Yukawa and T. Matsubara (eds.): *Solid State Physics I* (2nd edn) (Iwanami Series on Foundation of Contemporary Physics, Vol. 6, in Japanese) (Iwanami, Tokyo 1978).

Chapter 3 'Physics of Helium System' and Appendix A 'Superfluid ^{3}He' by the present author serve as moderate introductions.

[B-9] K. Yamada and T. Ohmi: *Superfluidity* (New Physics Series 28, in Japanese) (Baifukan, Tokyo 1995).

An excellent book on superfluid helium published recently.

Books on superconductivity

[C-1] P. de Gennes: *Superconductivity of Metals and Alloys* (Benjamin, New York 1966), (Reprint, Addison-Wesley, Reading 1989).

A unique textbook, which contains a detailed account of the Bogoliubov–de Gennes theory and the GL theory.

[C-2] R.D. Parks (ed.): *Superconductivity, Vols. I & II* (Marcel Dekker, New York 1969).

A handbook of research in superconductivity as of the end of the 1960s. Contributions to these volumes are repeatedly cited when one discusses such topics as the strong coupling theory.

[C-3] J.R. Schrieffer: *Theory of Superconductivity (revised edn)* (Addison-Wesley, Reading 1983).

An excellent textbook on the Green's function formulation of superconductivity and electron–phonon interaction.

[C-4] M. Tinkham: *Introduction to Superconductivity* (McGraw-Hill, New York 1975).

A well-presented textbook. A detailed account of the Josephson effect is given.

[C-5] S. Nakajima: *Introduction to Superconductivity* (New Physics Series 9, in Japanese) (Baifukan, Tokyo 1971).

Basic concepts are well explained. Although this book is meant to be an introduction, it is rather advanced.

[C-6] Japanese Physical Society (ed.): *Superconductivity* (in Japanese) (Maruzen, Tokyo 1979).

A collection of expositions from foundations to applications.

[C-7] P.F. Dahl: *Superconductivity* (American Institute of Physics, 1992).

A history from Onnes to HTCS.

Monographs on selected topics in superconductivity

[D-1] A. Barone and G. Paternò: *Physics and Applications of the Josephson Effect* (John Wiley, New York 1982).

This book can be recommended as a detailed exposition on the Josephson effect.

[D-2] G. Grimvall: *The Electron–Phonon Interaction in Metals* (North-Holland, Amsterdam 1981).

A detailed account of electron–phonon interaction in both normal and superconducting states.

[D-3] D.H. Douglass (ed.): *Superconductivity in d- and f-Band Metals* (Plenum Press, New York 1976).

[D-4] W. Buckel and W. Weber (eds.): *Superconductivity in d- and f-Band Metals* (Kernforschungszentrum, Karlsruhe 1982).

The above two books give detailed accounts of superconductivity in transition metals and intermetallic compounds.

[D-5] B. Deaver and J. Ruvald (eds.): *Advances in Superconductivity* (Plenum Press, New York 1983).

[D-6] D. Saint James, G. Sarma and E. Thomas: *Type II Superconductors* (Clarendon, Oxford 1969).

These are slightly old-fashioned textbooks on type 2 superconductors.

[D-7] D.N. Langenberg and A.I. Larkin (eds.): *Nonequilibrium Superconductivity* (North-Holland, Amsterdam 1986).
[D-8] R. Tidecks: *Nonequilibrium Phenomena in Superconductors* (Springer-Verlag, Berlin 1990).
The above two books explain such topics as TDGL, Carlson–Goldman mode and PSO in detail.
[D-9] V.L. Ginzburg and D.A. Kirzhnits (eds.): *High-Temperature Superconductivity* (English edn.) (Consultant Bureau, 1982).
We list only the above book on HTSC research before the discovery of copper oxide high-T_c superconductors.
[D-10] T. Ishiguro and K. Yamaji: *Organic Superconductors* (Springer-Verlag, Berlin 1990).
An up-to-date review on organic superconductors.
[D-11] J.C. Phillips: *Physics of High Temperature Superconductors* (Academic Press, London 1989).
This book provides a good explanation of the physics of high-T_c superconductors, although from a restricted point of view.
[D-12] Proceedings of the 19th International Conference on Low Temp. Physics, Part III, *Physica* **B169** (1991).
[D-13] S. Maekawa and M. Sato (eds.): *Physics of High-Temperature Superconductors* (Springer-Verlag, Berlin 1991).
[D-14] Proceedings of the International Conference on Materials and Mechanisms of Superconductivity, High Temperature Superconductors III, Part 1, *Physica* **C189** (1991).
The above three proceedings provided excellent coverage on the present status of research on high-T_c superconductors and heavy electron systems.
[D-15] K.S. Bedell, M. Inui and D.E. Meltzer (eds.): *Phenomenology & Applications of High Temperature Superconductors* (Addison-Wesley, Reading 1992).
[D-16] High Temperature Superconductors, special issue in *Solid State Physics* (in Japanese) **25** (1990) 617.
The above references contain convenient reviews for further, detailed study of HTSC.

Further reading

The following references are listed to supplement the main text and to cover those topics that are not considered therein.

[E-1] D.E. MacLaughlin: Magnetic Resonance in the Superconducting State, in *Solid State Physics* **31**, edited by H. Ehrenreich, F. Seitz and D. Turnbull (Academic Press, London 1976).

[E-2] A.J. Leggett, S. Chakravarty, A.T. Dorsey, M.P.A. Fisher, A. Garg and W. Zwerger: Dynamics of the Dissipative Two-state System, *Rev. Mod. Phys.* **59** (1987) 1.
[E-3] P.B. Allen and B. Mitrović: Theory of Superconducting T_c, in *Solid State Physics* **37**, edited by H. Ehrenreich, F. Seitz and D. Turnbull (Academic Press, London 1982).
[E-4] W.J. Skocpol and M. Tinkham: Fluctuations near Superconducting Phase Transitions, in *Report on Progress in Physics* **38** Part 3, (1975).
[E-5] M.M. Salomaa and G.E. Volovik: Quantized Vortices in Superfluid ^{3}He, *Rev. Mod. Phys.* **59** (1987) 533.
[E-6] A.I. Larkin and Yu.N. Ovchinikov: Pinning in Type II Superconductors, *J. Low. Temp. Phys.* **34** (1979) 409.
[E-7] G.E. Volovik and L.P. Gor'kov: Superconducting Classes in Heavy-Fermion Systems, *Sov. Phys. JETP* **61** (1985) 843.
[E-8] M. Sigirist and K. Ueda: Phenomenological Theory of Unconventional Superconductivity, *Rev. Mod. Phys.* **63** (1991) 239.
[E-9] M. Ozaki, K. Machida and T. Ohmi: Classification of superconducting pair function by symmetry, *Solid State Physics* (in Japanese) **23** (1988) 879.
[E-10] J. A. Sauls: The order parameter for the superconducting phases of UPt_3, *Adv. in Phys.* **43** (1994) 113.
[E-11] E. Hanamura: *Quantum Optics* (Springer-Verlag, Berlin, to be published).

Collections of original papers (Circulated only in Japan)

[F-1] Superconductivity, *Collected Papers in Physics* **153** (Japanese Physical Society, Tokyo 1966).
[F-2] Quantum Fluids, *Collected Papers in Physics*: New Edition **43** (Japanese Physical Society, Tokyo 1970).
[F-3] Superfluid ^{3}He, *Collected Papers in Physics* **190** (Japanese Physical Society, Tokyo 1975).
[F-4] Physical Properties of Organic Superconductors, *Collected Papers in Physics I* (Japanese Physical Society, Tokyo 1992).

References in text

[G-1] J.G. Bednorz and K.A. Müller: *Z. Phys.* **B 64** (1986) 189.
[G-2] H.F. Hess, R.B. Robinson, R.C. Dynes, J.M. Valles, Jr. and J.V. Waszczak: *Phys. Rev. Lett.* **62** (1989) 214.
[G-3] A.D.B. Woods and R.A. Cowley: *Canad. J. Phys.* **49** (1971) 177.

[G-4] V. Fock: *Z. für Physik*, **98** (1936) 145.
[G-5] N.E. Phillips: *Phys. Rev.* **114** (1959) 676.
[G-6] R.W. Morse and H.V. Bohm: *Phys. Rev.* **108** (1957) 1094.
[G-7] G.J. Dick and F. Reif: *Phys. Rev.* **181** (1969) 774.
[G-8] R.H. Hammond and G.M. Kelly: *Phys. Rev. Lett.* **18** (1967) 156.
[G-9] Y. Masuda and A.G. Redfield: *Phys. Rev.* **125** (1962) 159.
[G-10] L.C. Hebel and C.P. Slichter: *Phys. Rev.* **113** (1959) 1504.
[G-11] W.H. Parker, D.N. Langenberg, A. Denenstein and B.N. Taylor: *Phys. Rev.* **177** (1969) 639.
[G-12] J. Kwo and T.H. Geballe: *Phys. Rev.* **B23** (1981) 3230.
[G-13] R.P. Groff and R.D. Parks: *Phys. Rev.* **176** (1968) 567.
[G-14] A.A. Abrikosov: *Soviet Physics-JETP* **5** (1957) 1174.
[G-15] J.D. Meyer: *Appl. Phys.* **2** (1973) 303.
[G-16] R.J. Watts-Tobin, Y. Krahenbuhl and L. Kramer: *J. Low Temp. Phys.* **42** (1981) 459.
[G-17] W.P. Halperin, C.N. Archie, F.B. Rasmussen, T.A. Alvesalo and R.C. Richardson: *Phys. Rev.* **B13** (1976) 2124.
[G-18] J.E. Berthold, R.W. Giannetta, E.N. Smith and J.D. Reppy: *Phys. Rev. Lett.* **37** (1976) 1138.
[G-19] A.I. Ahonen, T.A. Alvesalo, M.T. Haikala, M. Krustus and M.A. Paalanen: *Phys. Lett.* **51A** (1975) 279.
[G-20] M.M. Salomaa and G.E. Volovik: *Rev. Mod. Phys.* **59** (1987) 533.
[G-21] V.P. Mineev, M.M. Salomaa and O.V. Lounasmaa: *Nature* **324** (1986) 333.
[G-22] K. Shimizu, N. Tamitani, N. Takeshita, M. Ishizuka, K. Amaya and S. Endo: *J. Phys. Soc. Jpn.* **61** (1992) 3853.
[G-23] W. Little: *Phys. Rev.* **134** (1964) A1416.
[G-24] A.W. Sleight, J.L. Gillson and P.E. Bierstedt: *Solid State Commun.* **17** (1975) 27.
[G-25] K. Kadowaki, J.N. Li and J.J.M. Franse: *Physica* **C170** (1990) 298.
[G-26] Y.J. Uemura: *Physica* **B169** (1991) 99.
[G-27] T. Hasegawa: *Proc.US-Japan Seminar on the Electronic Structure and Fermiology of HTSC*, 1992.
[G-28] Y. Kitaoka, K. Ishida, S. Ohsugi, K. Fujiwara and K. Asayama: *Physica* **C185-189** (1991) 98.
[G-29] J.N. Li, K. Kadowaki, M.J.V. Menken, A.A. Menovsky and J.J.M. Franse: *Physica* **C161** (1989) 313.
[G-30] H.R. Ott, H. Rudigier, T.M. Rice, K. Ueda, Z. Fisk and J.L. Smith: *Phys. Rev. Lett.* **52** (1984) 1915.

[G-31] Y. Kohori, T. Kohara, H. Shibai, Y. Oda, Y. Kitaoka, K. Asayama: *J. Phys. Soc. Jpn.* **57** (1988) 395.

[G-32] K. Hasselbach, L. Taillefer and J. Flouquet: *Phys. Rev. Lett.* **63** (1989) 93.

[G-33] H.Tou, Y. Kitaoka, K. Asayama *et al.*: *Phys. Rev. Lett.* **77** (1996) 1374.

References in appendices

[H-1] M.H. Anderson, J.R. Ensher, M.R. Matthews, C.E. Wieman and E.A. Cornell: *Science* **269** (1995) 198.

[H-2] K.B. Davis, M.-O. Mewes, M.R. Andrews *el al.*: *Phys. Rev. Lett.* **75** (1995) 3969.

[H-3] M.O. Mewes, M.R. Andrews, N.J. van Druten, D.M. Kurn, D.S. Durfee, and W. Ketterle: *Phys. Rev. Lett.* **77** (1996) 416.

[H-4] A. Griffin, D.W. Snoke and S. Stringari: *Bose–Einstein Condensation* (Cambridge University Press, Cambridge 1995).

[H-5] D.S. Jin, J.R. Ensher, M.R. Matthews, C.E. Wieman, E.A. Cornell: *Phys. Rev. Lett.* **77** (1996) 420.

[H-6] M.-O. Mewes, M.R. Andrews, N.J. van Druten, D.M. Kurn, D.S. Durfee and W. Ketterle: *Phys. Rev. Lett.* **77** (1996) 988.

[H-7] S. Stringari: *Phys. Rev. Lett.* **77** (1996) 2360.

[H-8] M.-O. Mewes, M.R. Andrews, D.M. Kurn, D.S. Durfee, C.G. Townsend and W. Kettlerle: *Phys. Rev. Lett.* **78** (1997) 582.

[H-9] D.A. Wollman, D.J. Van Harlingen, W.C. Lee, D.M. Ginsberg and A.J. Leggett: *Phys. Rev. Lett.* **71** (1993) 2134.

[H-10] J.H. Miller, Jr., Q.Y. Ying, Z.G. Zou *et al.*: *Phys. Rev. Lett.* **74** (1995) 2347.

[H-11] A. Mathai, Y. Gim, R.C. Black, A. Amar and F.C. Wellstood: *Phys. Rev. Lett.* **74** (1995) 4523.

[H-12] R. Chaudhari and S.-Y. Lin: *Phys. Rev. Lett.* **72** (1994) 1084.

[H-13] M. Sigirist, D.B. Bailey, R.B Laughlin: *Phys. Rev. Lett.* **74** (1995) 3249.

[H-14] E. Zeldov, D. Majer, M. Konczykowski, V.B. Geshkenbein, V.M. Vinokur and H. Shtrikman: *Nature* **375** (1995) 373.

[H-15] U. Welp, J.A. Fendrich, W.K. Kwok, G.W. Crabtree and B.W. Veal: *Phys. Rev. Lett.* **76** (1996) 4809.

[H-16] E.V. Antipov, S.N. Putilin, E.M. Kopnin *et al.*: *Physica* **C235-240** (1994) 21.

[H-17] Y. Maeno, H. Hashimoto, K. Yoshida *et al.*: *Nature* **372** (1994) 532.

[H-18] See, for example, M. Rapp, A. Murk, R. Semerad and W. Prusseit: *Phys. Rev. Lett.* **77** (1996) 928 and Yu.I. Latyshev and J.E. Nevelskaya: *Phys. Rev. Lett.* **77** (1996) 932.

It should be remarked that the above list is compiled with the author's prejudice and is not meant to exhaust all the important works.

Index